D1067970

Rural IMAGES

THE KENNETH NEBENZAHL, JR.,
LECTURES IN THE
HISTORY OF CARTOGRAPHY

PUBLISHED FOR THE
HERMON DUNLAP SMITH CENTER
FOR THE HISTORY OF CARTOGRAPHY
THE NEWBERRY LIBRARY

SERIES EDITOR, DAVID BUISSERET

THE UNIVERSITY OF CHICAGO PRESS
CHICAGO AND LONDON

EDITED BY

David Buisseret

Rural IMAGES

ESTATE MAPS IN THE OLD *and* NEW WORLDS

David Buisseret is Jenkins and Virginia Garrett Professor of History at the University of Texas, Arlington. He is author of *Historic Illinois from the Air* and editor of *Monarchs, Ministers, and Maps*, both published by the University of Chicago Press.

The University of Chicago Press, Chicago 60637
The University of Chicago Press, Ltd., London

Printed in the United States of America
05 04 03 02 01 00 99 98 97 96 1 2 3 4 5
ISBN: 0-226-07990-2 (cloth)

Library of Congress Cataloging-in-Publication Data

Rural images : estate maps in the Old and New Worlds / edited by David Buisseret.
p. cm. — (The Kenneth Nebenzahl, Jr., lectures in the history of cartography)
Includes bibliographical references (p. –) and index.
1. Real property—Maps—History. I. Buisseret, David. II. Series.
GA109.5.R87 1996
912′.09—dc20 95-30609
CIP

This book is printed on acid-free paper.

CONTENTS

ILLUSTRATIONS

EDITOR'S NOTE

The essays published here were for the most part first delivered in the ninth series of Kenneth Nebenzahl, Jr., Lectures in the History of Cartography, held at the Newberry Library, Chicago, in November 1988. It has taken a long time for the authors' work to reach publication, but that was not through any fault of theirs; when the essays had been assembled, it was plain that there were some serious gaps in the coverage, and I have down the years been doing my best to remedy this.

In the course of this work I have incurred many obligations. Coming to the study of North American large-scale mapping with a mind that was more or less a tabula rasa, I have received good counsel from many scholars and curators in the eastern and southeastern states. In New York, Professor Jo Mano spent many hours searching the local archives in a hunt for estate maps that eventually proved unfruitful—but nonetheless very helpful for understanding their incidence. I received assistance from Marcy Silver at the Maryland Historical Society, and from Howson Cole in the Virginia Historical Society; here too Allen Meyer, of the Chicago Map Society, made an exhaustive and instructive search for map types resembling estate maps.

From the start, Philip Morgan put me in touch with scholars working in the field, like Leland Ferguson of the University of South Carolina. In that state, territory par excellence for estate maps, I received much help from Cam Alexander and Mary Giles in the Fireproof Building (Historical Society of South Carolina) at Charleston, as well as from Frederick Holder of Pickens. In Louisiana, second only in interest to South Carolina for my theme, my guides were Bernard Lemann, Rose Lambert of the Louisiana State Museum, and the late John Mahé at the Historic New Orleans Collection.

In Germany, the Herzog August Bibliothek at Wolfenbüttel offered me its legendary hospitality, so that I could study large-scale German topographical mapping of early modern times; in Chicago, the Newberry Library has as always been an excellent working base, with colleagues like Robert Karrow and James Akerman always ready to correct my wilder excesses. I have also to thank the two anonymous readers of the University of Chicago Press, who made many helpful suggestions about the shape of the book, giving it much more coherence than it originally had.

Finally, it is a pleasure to acknowledge the generosity of Mr. and Mrs. Nebenzahl, without whose help the series of lectures would never have begun, and without whose continuing interest it would be much less thriving.

INTRODUCTION

DEFINING *the* ESTATE MAP

DAVID BUISSERET

Among cartographic types, the estate map occupies a rather peculiar place. It is possible, even if not desirable, to study most other types of map without reference to the social and economic system out of which they emerged. For instance, there have been good books written about early modern European printed world maps and eighteenth-century maritime charts, whose authors made only the slightest reference to sixteenth-century dynastic and economic structures, or to the demands of the ever-expanding European navies; the books hardly seem diminished by this.

Such a divorce between the map and its society is impossible in the case of estate maps, which insistently pose the problem, What social and economic circumstances gave rise to this map? It may well be for this reason that they have never been systematically studied, for it is difficult to relate cartographic and socioeconomic developments in a wide variety of countries. Another likely reason for their relative neglect is that they are hard to define and often difficult to reproduce: we have here been reduced to providing redrawings of some maps.

The two outstanding characteristics of estate maps are that they were commissioned by a private proprietor, whether an individual, like a South Carolina planter, or a corporation, like an Oxbridge college, and that they show that proprietor's "estate."[1] This is what distinguishes them from the large-scale state-sponsored topographical maps recently studied by Kain and Baigent in *The Cadastral Map in the Service of the State.*[2]

Of course, it is not always simple to distinguish between work commissioned by a sovereign state and that ordered by a large landowner, particularly in Germany of the ancien régime. However, this is not normally a very contentious area of definition; for instance, it is clear that when John Norden in 1607 drew the maps of James I's Windsor estate (figure 2.13), he was acting for James not as king but as a great landowner, and so the maps fall within our field. On the other hand, when the duke of Saxony in 1624 commissioned Wilhelm Schickhardt (1592–1635) to undertake a *Landesaufnahme* of the duchy, this was plainly a cadastral survey at the state level.

Generally speaking, estate maps were produced at a larger scale than these cadastral surveys. Their scale, which was nearly always marked upon them, was rarely less than 1:10,000 (a little over 6 inches to the mile), and could go much larger, permitting the surveyors to show not only the names of individual fields, but sometimes even the function of individual buildings. Thus the European estate plans often show what a farm's outbuildings were used for, and some West Indian examples allow us to follow the whole sequence of sugar production in a series of specialized structures. In general, because estate plans cover a small area at a large scale, they are remarkably accurate, especially in comparison with some of the contemporary small-scale printed maps.[3] This accuracy was partly achieved by the generally planimetric nature of estate plans. As we shall see, they sometimes include features shown diagonally, or in bird's-eye view, but as a type they are remarkable for the verticality of their presentation.

Since they were designed to serve a single patron, estate maps usually remained manuscript; there was no need to disseminate them widely in printed form. Sometimes, indeed, printed plans with much the appearance of estate plans were produced; for instance, one accompanies *L'Art de l'indigotier*, published at Paris in 1770 (figure 1).[4] Similarly, the printed maps of encumbered estates, designed to accompany the nineteenth-century sales of bankrupt British West Indian estates, somewhat resemble estate plans. But they were not designed for the essential purpose of the estate plan proper, which is to enable a proprietor better to manage an estate, both by maximizing rents and by assigning land to its best possible use.

This purpose was well explained by John Norden, who in *The Surveyors Dialogue* (London, 1618) remarked that the lord of many manors must "be aided by the skillful and industrious travail of some judicious surveyor, who finding by his view and examination the true values and yearly possibilities of his Lord's lands, may be a good meane to retaine his Lorde within compasse of his revenues." To this day many Caribbean estates retain nineteenth-century estate plans in their offices; they sometimes still serve to guide decisions about the routine of cultivation.

This aspect of the definition rules out the many plans made in the course of legal disputes over boundaries. It also excludes maps made primarily for technical purposes, such as the construction of a dike or a bridge. Of course, such maps often resemble estate plans in scale and general appearance, but we must be alert to their intended purpose if we are to keep our definition as tight as is necessary.

The detail and accuracy required in an estate map usually implied the employment of a trained surveyor, though on occasion a country gentleman might rise to the task. The word *surveyor* has an interesting and instructive etymology, reflected in the many meanings that it still possesses. Originally, a surveyor was a "supervisor," responsible for some aspect of overseeing and custodianship. In medieval Europe such oversight of a landed property generally involved the "naming and bounding" of it, the compiling—if necessary with the aid of twelve "good and lawful" men—of a written account of it. In the development described in this book, a map came to accompany and in some cases supplant this written account, and this map was the work of the old supervisor, or new surveyor. The process is reminiscent of a similar development in marine mapping, whereby the written *portolano* was supplemented and eventually sup-

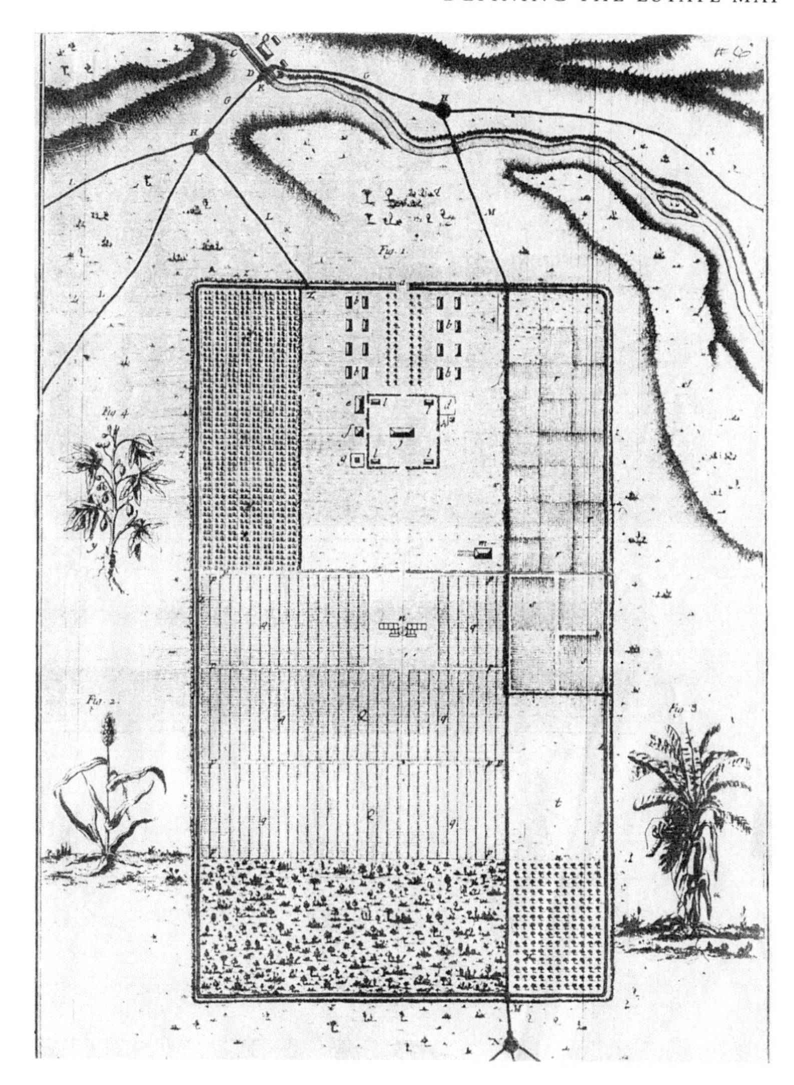

Fig. 1 N. de Beauvais-Raseau, plan of an indigo plantation, *L'Art de l'indigotier* (Paris, 1770)

planted by the portolan chart; again, a purely verbal descriptive form gave way over time to a more efficient cartographic one.

Ideally, the surveyor would be capable not only of producing an accurate plan, showing the estate or plantation in the detail needed to make decisions about its operation, but also of making it an elegant production, suitable for hanging in a conspicuous place. As William Leybourn put it in *The Compleat Surveyor* (London, 1653), "these things being well performed, your plot will be a neat Ornament for the Lord of the Mannor to hang in his study, or other private place, so that at pleasure he may see his land before him" (275).

An early example of an elegant North American estate plan comes from South Carolina (plate 4). Here the artist has to delineate an estate that is as yet little developed, but he shows a fine sense of style in setting it within a superb border, with an elegant coat of arms, an elaborate scale, and a finely wrought title cartouche. This is a North American counterpart of the magnificent plan that still hangs in the entrance of Long Melford Hall, in Suffolk, England. Composed in 1580 for Sir William Cordell by Israel Amyce, this remarkable monument uses nine sheets of parchment and measures over 8 feet by over 6 feet.[5] These maps were intended for public display, to confirm the wealth and extent of the owners' estates. They were in this sense the equivalent of engravings of "a gentleman's country seat" in eighteenth-century England, or of the oil paintings of plantation houses, invariably seen from the levee, so popular in eighteenth- and nineteenth-century Louisiana.

In general, estate maps are found only between about 1570 and 1900. Before 1570, as we shall see, the material conditions for the creation of such maps did not exist. After 1900 virtually all areas of large-scale agriculture were equipped with detailed printed maps, usually drawn up by the state, and these made the production of private maps unnecessary. We are dealing, then, with a transient cartographic form, brought into existence by particular circumstances of time and place.

Notes

1. One of the few attempts to define *estate maps* may be found in Helen Wallis and Arthur H. Robinson, eds., *Cartographical Innovations: An International Handbook of Mapping Terms to 1900* (London, 1987), 97–98. However, the editors do not entirely succeed in making a distinction between estate maps and the adjacent entries on cadastral maps and property maps. For the British Library's definition, see the letter of August 1967 by R. A. Skelton in *Cartographic Journal* 4, no. 1 (1967): 140.

2. Published at Chicago in 1992, this book won the Nebenzahl Prize.

3. See J. H. Andrews, *Plantation Acres: An Historical Study of the Irish Land Surveyor and His Maps* ([Belfast], 1985), 323, table 14.

4. See also the remarkable printed map of Wynnefield, made by Bernard Scalé in 1772 and reproduced in ibid., 335.

5. See Sarah Bendall, *Maps, Land, and Society: A History, with a Carto-Bibliography of Cambridgeshire Estate Maps, c. 1600–1836* (Cambridge, 1992), 25.

Chapter ONE

THE ESTATE MAP *in the* OLD WORLD

DAVID BUISSERET

England

The early history of English estate maps is described in chapter 2. Here we briefly consider some of the circumstances that brought them into being during the 1570s. Above all, the leaders of English society were becoming more aware of the potential uses of maps; as Peter Barber puts it, "By 1550, and still more so by 1580, . . . map consciousness—the ability to think cartographically and to prepare sketch maps as a means of illuminating problems—was becoming ever more widespread."[1]

As P. D. A. Harvey shows in chapter 2, cartographically alert landowners could call upon a long tradition of written surveys, and these could be transformed into maps through the efforts of surveyors, some of whom would have been familiar with ideas of scale from the fortification maps produced during the reign of Henry VIII. Others may have been inspired by the county maps produced (with the encouragement of Queen Elizabeth I) by Christopher Saxton during the 1570s. At all events, there seems to have been a pool of persons competent to make the maps now demanded by certain landowners.[2]

These landowners normally held substantial areas and were often concerned to maximize their rents, and perhaps to produce for the market rather than for mere subsistence; contrary to what one might expect, they were by no means all absentees. Much land had been changing hands in England since the confiscation of monastic lands by Henry VIII, and some maps may have been inspired in part by a desire to establish boundaries that were not as yet well known. It also seems important that England was enjoying a period of internal peace, between the tumults of the mid–sixteenth and mid–seventeenth centuries; surveyors obviously needed a certain tranquillity in which to operate. Finally, England in the sixteenth century had no general large-scale survey, of the kind that in many countries eventually made estate maps obsolete.

Spain

Spain provides a curious case. In its river valleys were vast estates, and at the highest levels its leaders in the sixteenth and seventeenth centuries were well aware of the potential of maps.[3] Moreover, of all the major countries of Europe in early modern times, Spain probably enjoyed the greatest degree of internal peace and security. However, it seems virtually certain that no estate maps were drawn in Spain before the eighteenth century, when some began to appear, no doubt under the influence of the French Bourbon administration.[4] Such large-scale topographical maps as did appear during the seventeenth century usually showed the areas around cities, perhaps reflecting the importance of the city in Spanish social and political life.[5]

How can we account for this early lack of estate maps? The major factor seems to be the absence in Spain of a sense of capitalist production. The great estates of the Guadalquivir and other valleys grew impressive quantities of agricultural produce, but this went to well-established traditional markets; there was little sense of any need to open up fresh outlets and so to increase production, using the land to greater effect. Moreover, for much of the sixteenth and seventeenth centuries the Spaniards were engaged in a vast overseas colonizing venture, which taxed their energies to the utmost and diverted them from a serious attempt to reassess the economic possibilities of the peninsula itself. Ironically enough, the highly map-conscious Philip II, who called for a vast series of maps showing his overseas possessions in considerable detail, never obtained similar coverage for his realms in the Old World;[6] it was as if they could be trusted to take care of themselves, in the time-honored fashion.

France

We find rather the same situation in France, where large estates in the many rich river valleys most often produced for traditional markets in the adjacent towns. But capitalist agricultural methods were slow to penetrate the French countryside, and the nation suffered from a devastating series of civil wars. Just when the estate plan was emerging in England in the 1570s, France was enduring the early years of the wars of religion, and the first half of the seventeenth century was equally marked by civil strife.

The French nobility, many of whose members by the middle of the seventeenth century had passed through the map-conscious Jesuit educational curriculum, were alert to the uses of cartography, but they generally commissioned detailed maps, not of arable farms, but of forests, mines, and property boundaries.[7] Marc Bloch observed as long ago as 1929 that there seemed to be no seigneurial "plans parcellaires" before 1650, and very few in the century after that.[8] He continued by noting that estate plans became much more numerous after 1740, attributing this to the so-called feudal reaction of the French nobility, who in many cases in the second half of the eighteenth century were taking their hereditary lands in hand, with a view to maximizing their output. Maps were an important part of this process; as the lawyer Joseph Rousselle observed in his *Instructions pour les seigneurs et leurs gens d'affaires* (Paris, 1770), "any land without a general survey and a geometrical plan is never well known in all its details."[9]

In the later eighteenth century, therefore, the French *arpenteurs* became very active, producing masterly large-scale maps showing the countryside in extraordinary detail.[10] Plate 7 shows one of 100 similar maps in an atlas by Christophe Verlet, drawn to show the parish of Busnes, near Lille in northern France. Each map shows one "canton," fully filling the page; as the cantons were of different sizes, the maps in the atlas have a wide range of scales and show a great mastery of proportion in composition. On the map of canton 34, it is possible to distinguish not only the houses and fields of the peasants, but also the house of the *curé*, his church, and "the lime-tree which is the place of publication of the parish of Busnes," where the king's edicts were read out. Other maps in this atlas show the roads, windmills, remnants of medieval strip fields, and so forth. Such atlases, which are relatively common for northeastern France, in effect allow us to reconstitute whole parishes in their spatial layout on the eve of the Revolution, and it is strange that they have not been more intensively studied.

The Low Countries

The Low Countries were in the early sixteenth century a region of great cartographic activity, both in the north and in the south. The south could boast the great cartographers of Antwerp and the mathematicians of the University of Louvain; in the north were thriving centers at Amsterdam, Haarlem, The Hague, and Middleburg.[11] There was also a particularly lively tradition of large-scale topographical mapping, associated often with diking and reclamation projects. These works were supervised by the polder authorities, or *waterschappen*, and the maps that they produced are known as *waterschapskarten*.[12] Later in the century these maps were often printed, for a large number of people were vitally concerned to know where the boundaries of the land ran.

In spite of this intense activity in dike maps, and the ubiquity of the *landmeter*, the Netherlands does not seem to have produced a large number of early estate plans. The *domanial kaart* is curiously lacking from the indexes and plates of books about the history of cartography in the Netherlands, at least before the eighteenth century.[13] The splendid series of maps of the estates of the former abbey of Rijnsburg, drawn in 1598 by Simon van Buningen, seems to be the exception rather than the rule. Perhaps what would have been the natural mapping role of great landowners in other countries was here taken over by the *waterschappen*, and their products were in many cases almost indistinguishable from the estate plans found in other regions.

One of the greatest landowners in the southern Netherlands were the ducs de Croy, and in the late sixteenth century Charles de Croy commissioned an extraordinary abundance of material showing his estates. The famous *Albums de Croy* contained about 2,400 marvelously crafted bird's-eye views of localities in France and what is now Belgium; they seem to be the work of Adrien de Montigny.[14] Plate 1 shows one of these views; the road from Menin to Lille traverses the foreground, by the village of Halluin, and a huge tree marks the crossroad. The image is full of life, with a cart on the road, boats on the river, and a windmill on the horizon; it is stylistically reminiscent of the paintings of the earlier miniaturists.

As well as these bird's-eye views, Charles also commissioned a few estate maps, probably the work of "Maistre Pierre Bersacques, ingéniaire et mesurer sermenté faisant les cartes, plantz, pourtraictz et mesureur des terres et seigneuries de Son Excellence."[15] These estate maps were models of their kind, but they plainly took second place in Charles's affections to the bird's-eye views, which are more elegant and excellently served the purpose of celebrating the extent and beauty of his possessions. It seems that Pierre Bersacques did not work for other patrons, and it is not easy to find other published examples of early estate maps in the southern Netherlands.

The German-Speaking Regions

The German-speaking regions were astonishingly active in mapping about 1500. This cartographic activity rested on a deep tradition of mathematical and astronomical knowledge, available chiefly through the universities. As early as the mid–fifteenth century, compasses had been produced in Augsburg, Nuremberg and Munich, and the first German observatory had been founded at Nuremberg in 1475.[16] Moreover, the German presses were technically very advanced, so that the Spanish and Portuguese maps of the New World usually were printed in Germany, by editors like Waldseemüller and Münster. The first earth globe seems to have been made in Nuremberg in 1492, and that city was for years the center of European globe production.[17] After the middle of the century, when Abraham Ortelius was collecting material for his great atlas, the *Theatrum orbis terrarum*, published in 1570, the contributions from the German-speaking areas were outstanding in their quantity and quality.[18]

This extensive cartographic activity was reflected in the early appearance of manuals of instruction for topographical mapmakers.[19] Albrecht Dürer was convinced of the importance of what he called *Messung*, that is, the mastery of those mathematical principles that would allow all artists—and surveyors—to do their best work. In the *Underweysung der Messung*, published in 1525, he set out the principles through which a true rendering of nature could be achieved.[20] Dürer, of course, was an accomplished mathematician, but the manuals published for surveyors in Germany fell far short of his elegant theories. One of the earliest was the *Geometrei* of Jacob Köbel, first published at Frankfurt in 1531 and reprinted eight times by 1616.

First, he writes, you must "take sixteen men, large and small," as they come out of church, and line them up with one foot upon a rod (figure 1.1). This rod becomes the surveyor's 16-*Schuh* staff, with which fields of various sizes can be measured.

Geometrei / Von künstlichem
Messen vnd absehen / allerhand höhe / fleche /
ebene / weite vnd breyte / Als Thürn / Kirchen /
bäw / baum / velder vnd äcker ꝛc. mit künstlich zübereyten Jacob stab / Philosophischen Spiegel / Schatten / vnd Meßrüten / Durch schöne Figurn vnd Exempel / Von dem vil erfarnen H. Jacob Köbel / weiland Stattschreiber zu Oppenheim / verlassen.

Fig. 1.1 Jacob Köbel, sixteen men making up a surveyor's rod, *Geometrei: Von künstlichen Messen* (Frankfurt, 1536)

Here we recognize the English "rod, pole, or perch," normally cited as 16.5 feet long, and a standard measure in England for many years. In contrast with the English writers who began to appear later in the century, Köbel does not write as if for a lord desiring to know the bounds of his property, but as if for a peasant-farmer anxious to avoid squabbles with his neighbors; we might call his book a do-it-yourself manual for smallholders, which no doubt explains its great popularity.[21]

Other manuals appeared down the years, for example, Johannes Stöffler, *Von künstlicher Abmessung* (Frankfurt, 1536), Christoff Pühler, *Geometria oder Feldmessung* (1563), and Erasmus Reinhold, *Bericht vom Feldmessen* (Erfurt, 1574). On the whole, they do not represent much technical advance on Köbel's work. As late as 1578, Herman Widekind published at Heidelberg his *Bewirte Feldmessung und Theilung,* which in its tone much resembles Köbel. Addressing himself directly to the husbandman, he begins, "Herman Widekind wunschet dem Ackerman heil." Much of his work is taken up in defining various measures, but he does advocate, more explicitly than Köbel, the use of the compass.

Twenty years later, Paul Pfinzing published his *Methodus geometrica* at Nuremberg.[22] Previous survey manuals had been small format and rather inelegantly printed, but Pfinzing's was a fine large book of about ninety pages, with many plates. He very explicitly set out the method of survey, by taking measured compass bearings and then transferring these to paper (figure 1.2). In another plate, he shows the surveyor on a high piece of ground, measuring the angles to various adjacent villages; he did not go so far as to advocate triangulation, though this technique was well known in many learned German circles, including those among which he moved at Nuremberg. Pfinzing calls his object the production of a "Landtafel oder Mappam," and gives us an example of what the finished work ought to look like (figure 1.3). It is in effect a bird's-eye view of the landscape, with orientation but of course without scale; as Pfinzing puts it, we see the land here "as if in a mirror." All the houses and natural features seem to be included, but there is no attempt accurately to measure woods or fields.

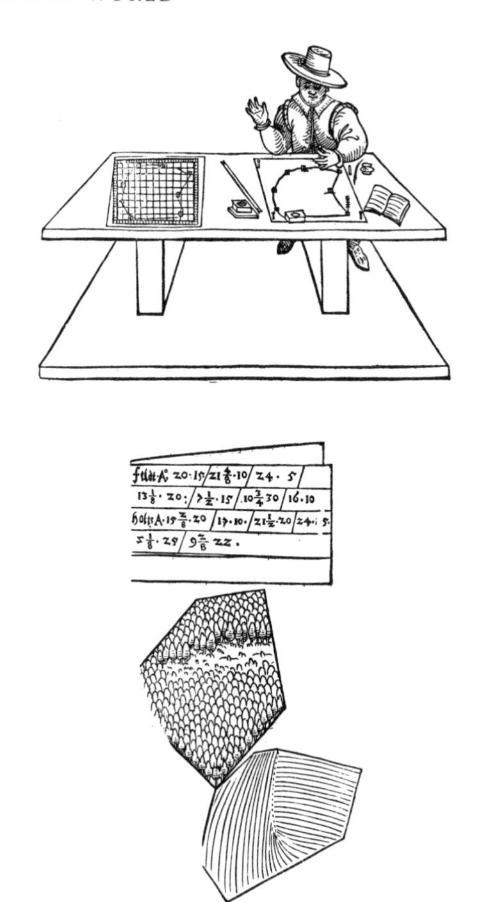

Fig. 1.2 Paul Pfinzing, the method of making a survey, *Methodus geometrica* (Nuremberg, 1598) (Herzog August Bibliothek)

Pfinzing was not, indeed, much concerned about a high degree of precision. For measurement he advocated various devices, including counting

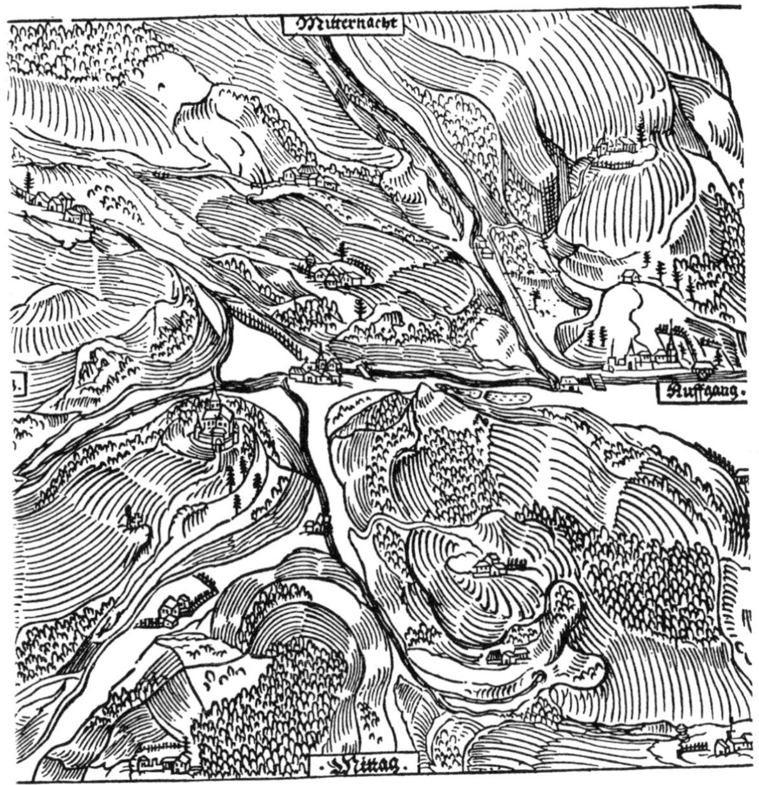

Fig. 1.3 Paul Pfinzing, an ideal field plan, *Methodus geometrica* (Nuremberg, 1598) (Herzog August Bibliothek)

the number of paces a horse made, or even a mechanical measuring cart (figure 1.4); as he says of this, while riding at his ease in it, "one finds measuring from a cart the best way." For measuring irregular fields, he suggests the ingenious expedient of cutting out a paper map of the field, and then weighing it against a paper cut-out of known size (figure 1.5). All in all, we are not surprised by his remark that "land ought not to be measured out as carefully as saffron."

Pfinzing's work is among those cited by Levinus Hulsius, whose *Erster Tractat des mechanischen Instrumenten* was published at Frankfurt in 1603. After listing his sources, Levinus gives first a geometrical and then an arithmetical explanation of surveying operations. He then describes the "planimetra" used to measure fields and buildings. This is in effect an early theodolite; in these years such instruments were coming into use all over Europe.[23] Another contemporary device was the plane table, illustrated and described in Daniel Schwenter's *Geometriae practicae novae*, published at Nuremberg in 1617 and often reprinted. Schwenter's four books fully explain how to measure a variety of geometrical shapes, and there can be no doubt that he would have been capable of producing a planimetric map, with all the elements viewed vertically. When he came to provide his readers with a model map, however, he chose the traditional method of the bird's-eye view (figure 1.6). Like

Pfinzing, he wanted to show the land "as if in a mirror," no doubt because he felt that in this way it could be more easily apprehended by his readers.[24]

By the early seventeenth century, then, a wide variety of manuals were available in Germany for people interested in surveying. These manuals had been published in a number of cities, and the circulation of books was such that many must have been widely available. Unlike the contemporary publications in England, they remained rather "uncapitalist" in tone; their authors were not interested in serving landlords who were anxious to improve the land (and their tenants). On the contrary, Köbel seems mainly interested in keeping the peace between neighbors, and the later writers often have a markedly academic tone, shown in their very long geometrical preambles. German patrons were interested in knowing the boundaries of their land, but they did not view the map as a means of more effectively exploiting their possessions.

Fig. 1.4 Paul Pfinzing, the author riding in his measuring cart, *Methodus geometrica* (Nuremberg, 1598) (Herzog August Bibliothek)

Topographical mapping in Lower Saxony and Oldenburg

Lower Saxony was a region that had a particularly strong tradition of topographical mapping, in part no doubt because much of its area consisted of rich farmland. It was also a region of learned and

Fig. 1.5 Paul Pfinzing, the author weighing fields, *Methodus geometrica* (Nuremberg, 1598) (Herzog August Bibliothek)

Fig. 1.6 Daniel Schwenter, an ideal field plan, *Geometriae practicae novae* (Nuremberg, 1617) (Herzog August Bibliothek)

wealthy princes, in particular the dukes of Braunschweig-Lüneburg at Celle, and of Braunschweig-Wolfenbüttel at Wolfenbüttel, and the margraves of Oldenburg. From this area come many examples of maps on a scale slightly smaller than the estate map, known in German as *Landesaufnahmen*. For instance, the cartographer in Celle about 1600 was Dr. Johann Mellinger, who had provided the map of Thuringia for Ortelius. In the service of the duke of Braunschweig-Lüneburg, Mellinger produced the celebrated *Ämteratlas des Fürstentums Lüneburg*, which consisted of a key map and forty-two detailed maps of individual *Ämter*. These territorial maps, at a scale of about 1:80,000, were remarkable for their elegance, and were copied for many years.[25]

Mellinger's counterparts at Wolfenbüttel were Johann Krabbe and Johann Tiele, both appointed in 1586 as surveyors to Duke Heinrich Julius.[26] We know a good deal about Krabbe, who attended the Universities of Helmstedt and Frankfurt an der Oder, becoming acquainted with the leading mathematicians. By 1579 he had made his first astrolabe, and from 1586 onward he collaborated with Tiele in the service of the duke. Their talents were complementary, for while both were instrument makers, architects, and surveyors, Tiele was in addition a millwright and master of fireworks; Krabbe excelled in the more academic areas of music, geography, and astronomy.

Some of their maps survive in the Staatsarchiv of Hannover and Wolfenbüttel, demonstrating a wide range of expertise. Krabbe drew in 1591 a very remarkable *Chorographia der Hildesheimischen Stiftsheide,* to show the ravages of the war of 1519. Centered on Hannover, it takes in Minden to the west and Wolfenbüttel to the east, at a scale of about 1:125,000. Superbly reproduced in *Niedersächsen in alter Karten* (n. 19), it is the work of a master. In 1603 Krabbe composed a fine forest map, the *Abriss des Sollings,* and over the years drew some field plans, which are remarkable for the elegance of their calligraphy. Ernst Pitz, who has made the closest study of the work of Krabbe and Tiele, claims that some of their work came about as a result of the dukes' desire to farm their land more efficiently.[27] This is not obvious from their maps, but it is evident that they were accomplished surveyors, able to work at a wide variety of scales. Probably they did have an idea of the way in which their maps enabled the duchy to be more efficiently managed, but neither they, nor any other surveyors in the area, show anything like the conscious urge toward land improvement of the English surveyors of the late sixteenth century.

Tiele's and Krabbe's counterpart to the north, in Oldenburg, was Johann Conrad Musculus, probable author of the very elegant *Karte der Grafschaft Oldenburg,* produced in 1641.[28] Musculus, whose family name was Mauskopf, was born in 1587 in Strassburg, the son of a bookbinder. He moved to Oldenburg and by 1621 was drawing maps, though we have no idea how he was trained. In 1629 he became *Wallmeister* for Oldenburg, responsible for the maintenance of the very extensive walls and outworks; many of his maps were drawn in connection with his post.

As figure 1.7 shows, his work was characterized by great sobriety and precision. In his masterly *Niedersächsen in alter Karten* (81), Heiko Leerhof reproduces this map alongside a modern map of the same area, and it is easy to appreciate the accuracy of Musculus's work. Writing in 1937, Gustav Ruethning went so far as to claim that Musculus had "been the pioneer of cadastral mapping in Oldenburg,"[29] and this may not be far from the truth. Musculus and the other cartographers of Oldenburg were no doubt influenced by the maps made not so far away in the United Provinces, where the need to protect land from water led to the drawing of elegant and precise large-scale maps, akin to estate maps.

Mapping in Rhineland/Westphalia, Hesse, and Württemberg

The southwest region's early maps are particularly well covered in *Geschichte in Karten: Historische Ansichten aus den Rheinland und Westfalen.*[30] If these examples are representative, however, there was very little planimetric field measurement before the eighteenth century, when it became common. One of the few exceptions is the work of Hermann tom Ring (1521–97), a Münster-based painter whose field plan, at a scale of 1:4,000, resembles an estate plan (figure 1.8); he seems to have been the only one working at that scale.

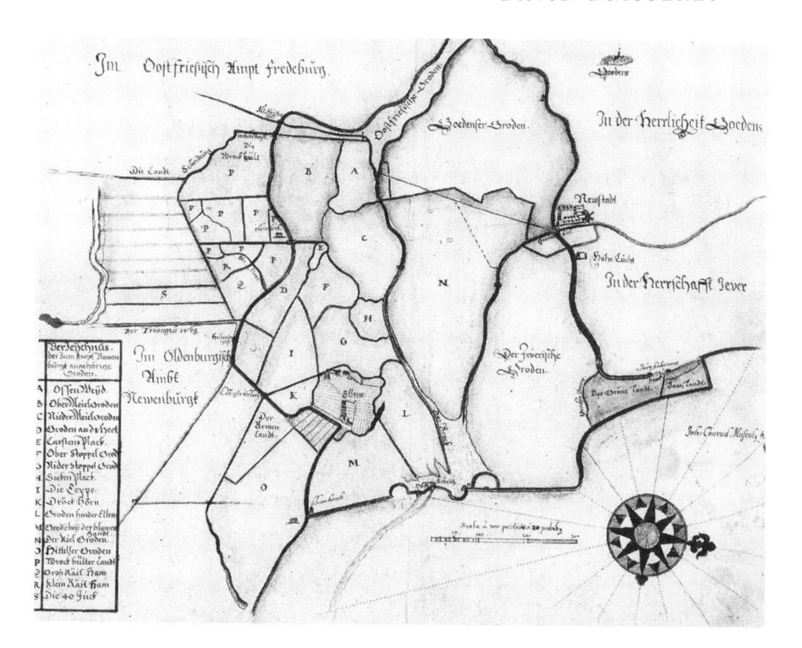

Fig. 1.7 Johann Conrad Musculus, plan of the old and new polders in Amt Neuenburg, 1635 (Niedersächsisches Staatsarchiv Oldenburg Best. 298 MN No. 1)

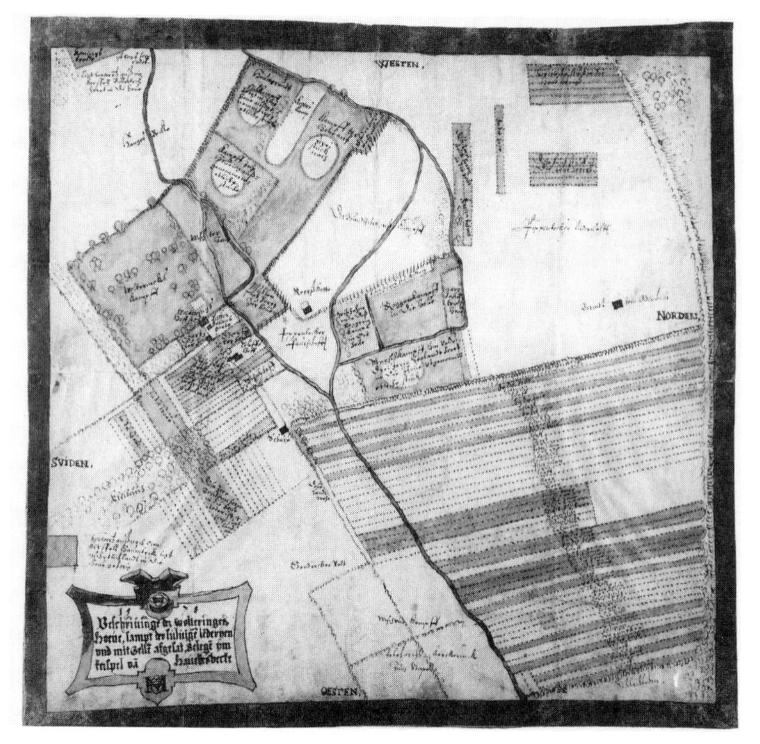

Fig. 1.8 Herman tom Ring, field plan of Eerbes Woltering, 1578 (Nordrhein-Westfälisches Staatsarchiv Münster, Map Collection A 2387)

A little to the northwest lay Hesse, where between 1607 and 1621 Wilhelm Dilich worked for Landgrave Moritz. Born about 1571, Dilich trained at the Universities of Wittenberg and Leipzig, and became known in the 1590s for his city plans. Landgrave Moritz commissioned him for a vast project that involved mapping Hesse with twelve general maps and fifty-eight *Ämter-Karten*. Eventually Dilich completed thirty maps and thirty-five town plans, at scales varying from 1:5,000 to 1:30,000.[31] His maps were not only very beautiful, but also relatively accurate.[32] At the largest scale, as in plate 2, they show much detail. This example, *Schloss und Bezirk Kleckenbühl*, is at a scale of 1:5,400, allowing Dilich to show the moated castle (drawn in more detail in an accompanying plan) with its buildings and fields listed from *A* to *P*, and the surrounding woods (of two types) and fields. In the inscription at the bottom of the map, Dilich notes that he drew it "using geometrical instruments" during 1621 and 1622. An elevation of the castle completes this exquisite work, which seems to have no counterparts in Hesse.

South of the Rhineland and Hesse lies Württemberg, intensively mapped during the sixteenth and seventeenth centuries. The earliest portrayal was by Heinrich Schweikher (1526–79), in a manuscript atlas now in the Württembergischen Landesbibliothek in Stuttgart.[33] Schweikher was born into an influential family in Sulz, and at a young age became a town official. In 1567 the Synod of Württemberg appointed him as one of two *Waisenvogte*, or protectors of orphans, and in this capacity he travelled extensively in the area around Stuttgart.

These travels no doubt gave him the opportunity to know the land that he mapped. We do not know why he undertook the work, or indeed what his qualifications were. But he did produce a general map and fifty *Ämter* maps. They have been judged to be relatively inaccurate[34] and certainly look a little amateurish; for instance, he made no attempt to relate the individual maps to the general map, as Apian had done so successfully in Bavaria in 1568. Their scale is also quite small, about 1:150,000. But they no doubt served their purpose, which was to give the synod some idea of where the villages around Stuttgart were.

While Schweikher was producing this work, Georg Gadner (1522–1602) was working on a series of maps of the duchy of Württemberg.[35] He was born in Bavaria and trained at Ingolstadt but, having converted to Protestantism about 1550, had to leave his country, and entered the service of Duke Christoph of Württemberg. He began drawing maps about 1560, and in 1572 completed his first map of the duchy; the original is lost, but Ortelius used it for the 1579 edition of the *Theatrum orbis terrarum*. In the course of compiling this map, Schweikher must have made sketches of sections of the country, and it is probably from one of these that he derived *Das Stuttgartner Amt* of 1589.[36] It is an elegant work, at a scale of about 1:16,000, conceived as a bird's-eye view.

As Oehme points out, the work of cartographers like Gadner represents an intermediate step between provincial maps and the planimetric topographic map.[37] Gadner's work was carried on by Johannes Oettinger (1577–1633), a literary figure whose maps are very similar to those of Gadner.[38] At the larger scale the Württemberg landscape continued to be delineated in the traditional way, as in the very accomplished *Landtafel* of 1602 by Philip Gretter (figure 1.9). This type of bird's-eye view had been recommended by the writers of surveying manuals, as we have seen, and continued to be the map of choice for showing localities.

The work of Gadner and Oettinger was carried on by the two Schickhardts, Heinrich (1558–1635)

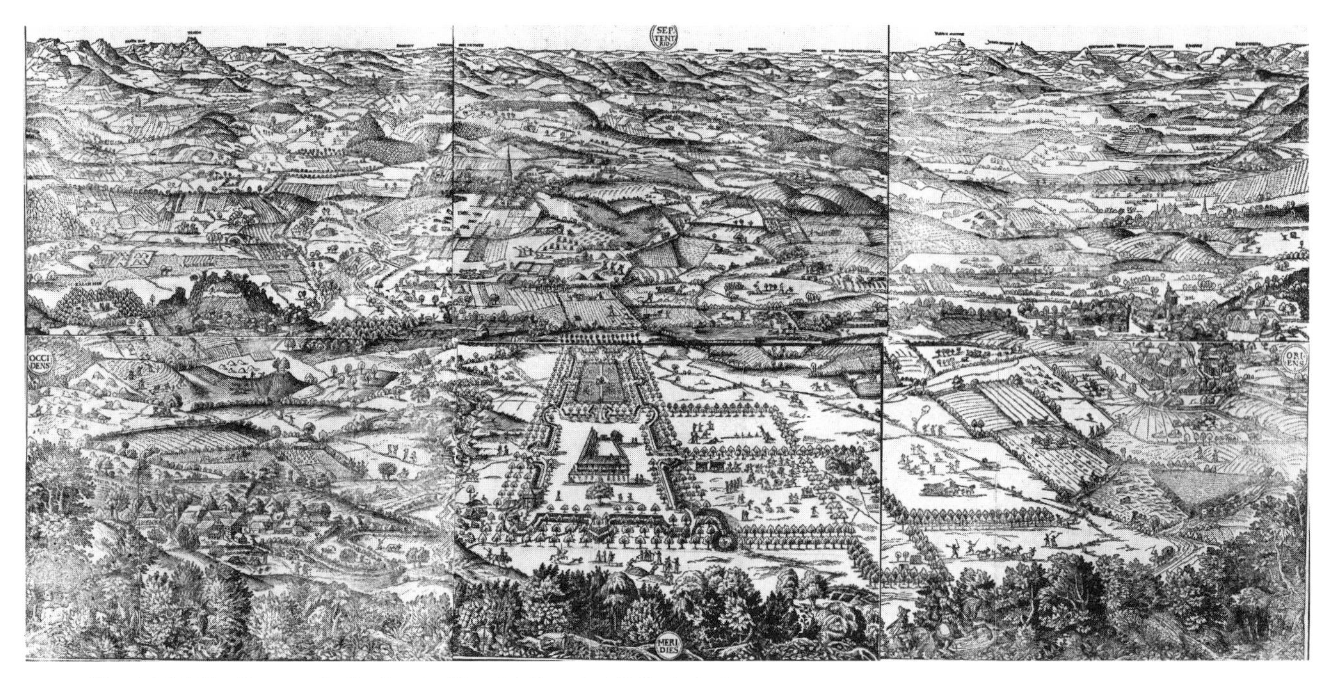

Fig. 1.9 Philip Gretter, *Boller Landtafel,* 1602 (Landesbibliothek, Stuttgart)

and his nephew Wilhelm (1592–1635).[39] Heinrich, the son of a woodcut maker, became a painter, fortress designer, and cartographer; he is best known for his *Aufnahme der Grafschaft Mömpelgard,* but produced many other maps at scales of 1:56,000 and smaller. Wilhelm Schickardt was an even greater polymath: theologian, orientalist, astronomer, and mathematician. From 1624 onward he worked on a great territorial survey of the duchy, using triangulation points to achieve a high degree of accuracy.[40] He explained his methods in the *Kurze Anweisung wie künstliche Landtafeln aus rechtem Grund zu machen,* published at Tübingen after his death, in 1669. The scale of his one completed map for the great project was small, about 1:130,000;[41] alas, the work was interrupted when he and his family died of the plague in 1635.

Topographic Mapping in Bavaria

The basis for the mapping of Bavaria was laid by the remarkable map commissioned from Philip Apian and published in 1568.[42] First drafted at a scale of 1:45,000, it was eventually printed at 1:135,000, so that Apian was able to include the names of villages and many natural features. His work came in twenty-five sheets, of which the first was a key map, showing the whole of Bavaria divided into twenty-four sections, each of which corresponded to one of the following maps. Apian's work was as accurate as it was innovative, for the whole structure had been astronomically determined, and the ground distances established by triangulation and compass bearing. The set of maps was used up to the nineteenth century, and may indeed have deterred others from attempting to map the same area.

The most interesting maps of the Nuremberg region were drawn by Paul Pfinzing, author of the *Methodus geometrica* of 1598. Son of a patrician Nuremberg family, he attended the University of Leipzig, and perhaps travelled in the Netherlands and in France.[43] Returning to Nuremberg, he held a succession of posts in city government, where he was conspicuous for his use of maps in such areas as planning for flood control and contracting for the repair of walls. Over fifty of his maps survive, thirty-four of them in the *Pfinzing-Atlas* preserved at the Staatsarchiv in Nuremberg. His maps of the countryside cover a wide range of styles. Figure 1.10 for instance, shows his home village of Hennenfeld very much in the picture style of his ideal map (figure 1.3). He drew this in 1585, orienting the map to the south and labelling many of the fields; other features are indicated in a key numbered from one to nine. The scale is about 1:15,000, and this map is in essence a bird's-eye view.

Pfinzing could also compile maps that were planimetric. He drew a strictly planimetric plan of Nuremberg, and some of his later *Amt* plans lean in this direction. For instance, in 1592 he made another plan of Hennenfeld that differs sharply from the plan of 1585 (figure 1.11). This time there is a scale (bottom left), and a much fuller list of fields and of people's holdings, running from one to seventy-nine. The scale is now about 1:3,840; this map could have been used to make decisions about

Fig. 1.10 Paul Pfinzing, *Landtafel* of Hennenfeld, 1585 (Staatsarchiv, Nuremberg, *Pfinzing-Atlas*)

Fig. 1.11 Paul Pfinzing, plan of Hennenfeld, 1592 (Staatsarchiv, Nuremberg, *Pfinzing-Atlas*)

how the village and its fields were to be run—an estate map.

Such maps are exceedingly rare in sixteenth-century Bavaria. At the end of the century some were drawn by Peter Zweidler, on behalf of Bishop Neidhard von Thungen.[44] Zweidler came from Teuschnitz, and seems to have worked in Nuremberg as a bookbinder. When he came to serve the bishop in the Hochstift Bamberg, it was as "Fürstlich Bambergischer Landabreisser," or "Chorograph und Landmesser." Between 1597 and 1608 he produced twenty-five maps, many of which show forested country at a relatively large scale. His map of the Oberamt Vilseck is characteristic of his style (figure 1.12). Villages and roads are carefully drawn, and so are many fishponds along the rivers. Note on the right-hand side the small black stones that mark the edge of the property. There seem to be no studies on the accuracy of Zweidler's maps, but he certainly gives an impression of care and precision; however, he was probably unique in adopting this kind of map at this period.

Upper Saxony

In Upper Saxony the Universities of Leipzig and Wittenberg played an important part in teaching mathematical theory. As we have seen, the survey manual of Erasmus Reinhold, *Bericht vom Feldmessen*, was published at Erfurt in 1574, and successive dukes encouraged scientific activity. Thus Duke August (1526–1586, duke from 1533) founded the Dresdener Kunstkammer, and Duke Christian I (1560–1591, duke from 1586) carried on his interest in mathematics, geometry, and scientific instruments.[45]

The Ortelian map of Saxony was the work of Johannes Criginger (1521–71), who studied at the University of Wittenberg, Leipzig, and Tübingen, and seems to have worked without a patron. Meanwhile Duke August had commissioned the Leipzig professor Johann Humelius to work on a general

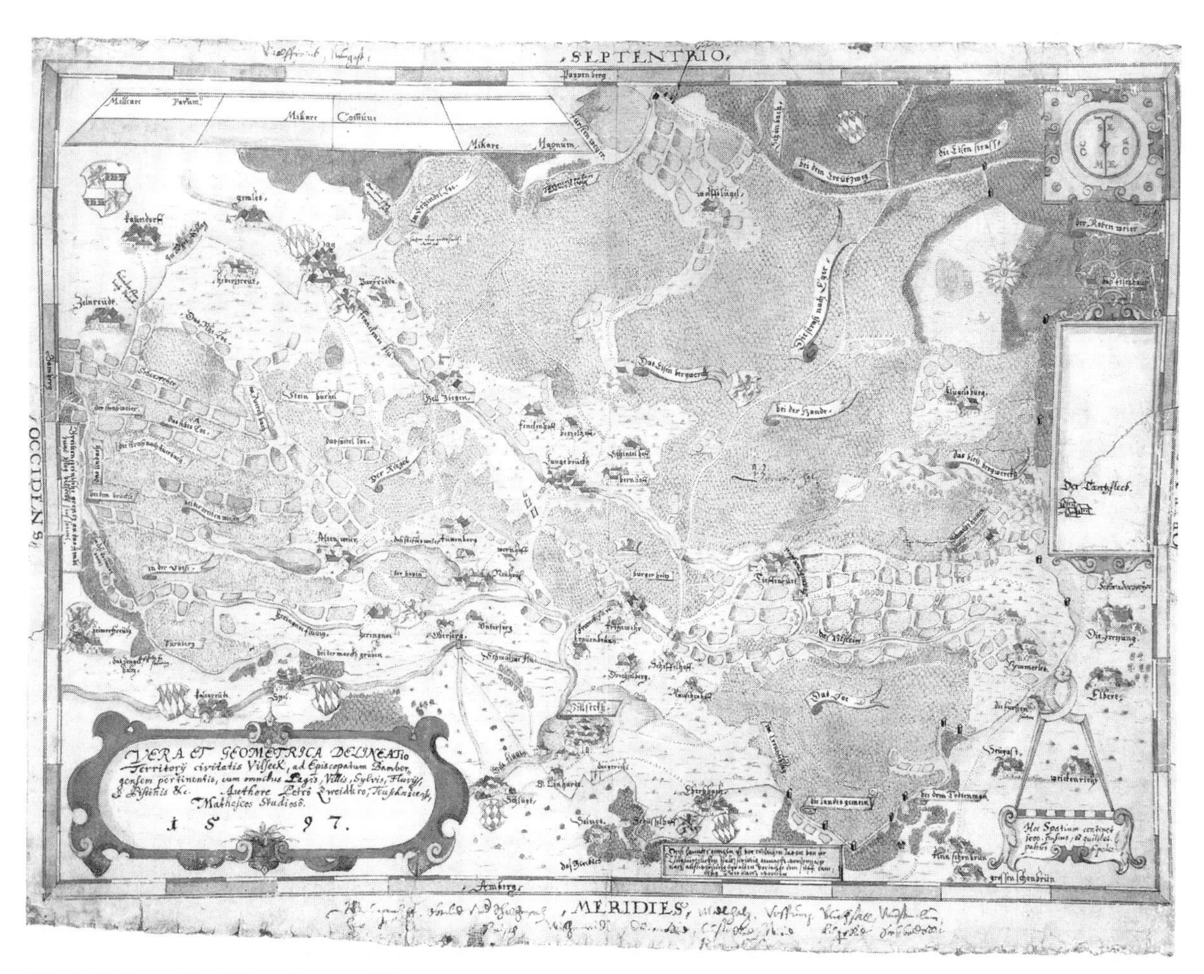

Fig. 1.12 Peter Zweidler, Oberamt Vilseck, 1597 (Staatsarchiv, Bamberg)

survey; not much was accomplished between 1555 and his death in 1562. Humelius was succeeded by Georg Oeder, who worked for the duke until 1575, making building surveys, sketches of heaths and woods, and finally a series of sixteen forest surveys. Oeder had the promising title of "Kurfürstlichen Sächsischer Markscheider," but it is hard to find reproductions of his work. He was succeeded in 1575 by his younger brother Matthias Oeder, who was commissioned to make a "general Land-Mappen." He worked on it between 1586 and 1607, eventually producing a masterwork at a scale of about 1:13,000, which allowed him to include such features as mills and farms. Matthias Oeder was an extraordinarily conscientious surveyor, whose work attained a high degree of precision.[46] After his death came the Thirty Years' War, which meant that in Upper Saxony no more detailed territorial mapping would be undertaken before the eighteenth century.

Switzerland

Johann Rauch probably came from the Vorarlberg region, but we know nothing about his birth or training.[47] In 1601 he went to the old imperial city of Wangen, where he worked as a house and picture painter. From 1608 to 1628 he also painted about twenty *Landtafeln* of towns and estates on the eastern shore of the Bodensee and in Upper Swabia. His work is remarkably accurate,[48] and remarkably beautiful, as may be seen from his *Landtafel* of Wangen; painted in 1616, it is still in the Heimatmuseum there. Such a map would have been constructed by using a chain and compass to compose a framework of polygonal figures, and then filling in the details by pacing and estimation.

Between 1626 and 1629 Rauch worked for the municipal council of Lindau on a *Landtafel* of the area under the council's jurisdiction. In the course of this work, however, it became apparent that some areas would not be delineated with sufficient detail, and so in April 1628 the Lindau municipal council ordered Rauch to "tear out and draw anew the four villages of Rickenbach, Aschach, Schonau and Oberreitnau, on paper, only roughly done [?], add the names, and send them as soon as possible"[49]

Rauch duly obeyed this order, and his map of Rickenbach has come down to us (figure 1.13). Oriented northward, it has a scale of about 1:2,750 and is drawn from so high an angle that it is virtually a planimetric map. Each house is marked by a number and the owner's name; fields are also identified. With a map like this, the Lindau council could much more easily control its estates than with Rauch's normal *Landtafel.* It is interesting that Rauch could apparently draw such maps upon demand, but that he clearly thought that most of his clients would be better satisfied with the more easily read *Landtafeln.*

The German-speaking areas: Conclusion

It is clear from this survey that the expertise existed in sixteenth-century Germany for the drawing of detailed, planimetric topographical maps. Cartographers like Pfinzing, Krabbe, Musculus, Dilich, and Rauch were quite capable upon occasion of executing maps of this kind. Such maps constitute a very small part of the total cartographic production,[50] however, and could in no way be described as a rich source. By and large, Germany was mapped with two cartographic types, the *Landtafel* at a local level, and the *Landesaufnahme* at the provincial level, with nothing in between.

Why did detailed planimetric maps of the German countryside not emerge before the eighteenth

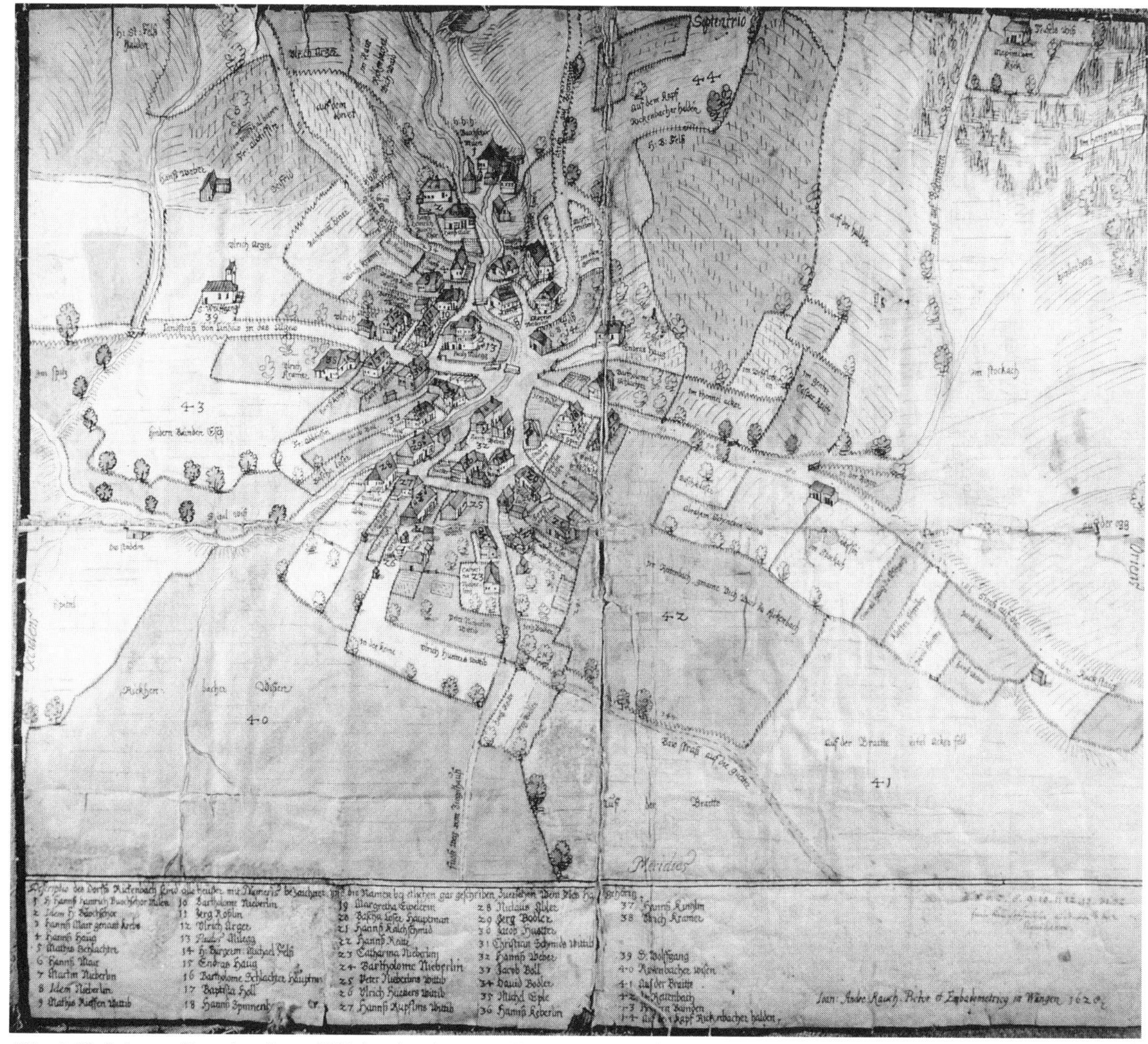

Fig. 1.13 Johann Rauch, plan of Rickenbach, 1628 (Landesregierung Archiv Tyrol, Innsbruck)

century, when they became quite numerous? The main reason must surely be that the *Landtafel* was well established as a cartographic form. It not only was familiar and easy to read but also fell into a recognizable category; it was the way to show a land of fields and forests, just as other specialized map types showed mining operations or seaways.

It may also be that the land market in Germany was relatively sluggish, and German agriculture slow to be penetrated by capitalist ideas. Political divisions too may have had an effect. In Germany political authority and effective management lay in larger subdivisions than in England, where the estate plan most conspicuously flourished, and these larger subdivisions were best shown by smaller-scale maps. Finally, politics in the wider sense had an influence on the development of German cartography. Whatever tendencies toward intensive large-scale mapping may have been developing in the early seventeenth century were effectively stifled after the outbreak of the Thirty Years' War in 1618, when so much of the country fell into ruin. Thus it was not until the eighteenth century that a majority of large German landlords began to use the planimetric estate map as a tool of management.

This survey has covered the region west of the Elbe, leaving out the vast servile estates of Brandenburg, Pomerania, and the other eastern provinces. Here one might have expected to find estate maps; after all, the units were large, and they produced huge quantities of cereals for an external market. However, none seems to exist; perhaps the Junker were lacking in map consciousness, or perhaps the land market was excessively static. It may be, too, that they were actually hostile to the idea of surveys, which in the hands of the sovereign might be used to encroach upon their constitutional rights.[51] For whatever reason, the eastern regions were mapped only at the small scale of the remarkable work of Casper Hennenberger.[52]

Italy

In principle, we should expect to find early estate plans for Italy, for the great centers like Venice, Florence, Milan, Rome, and Naples had leaders who commissioned a variety of plans for different purposes,[53] and there was no lack of trained surveyors. However, it has not been possible so far to identify an abundance of such plans in the literature.

Works on Liguria, Romagna, and Naples reproduce either no estate plans or only eighteenth-century ones. In Liguria there were fine early city views and plans of fortifications along the coast; perhaps the hinterland was too hilly, and the farms too small, to encourage the emergence of estate maps.[54] None appear to have been drawn in Romagna, either, though here the rich coastal plain might have been a propitious area for estate plans.[55] In Naples they first appear in the eighteenth century, under the influence of landowners who were probably familiar with agricultural practices in northern Europe.[56] In Baricchi's work on the region of Reggio Emilia a good many estate maps are shown, but almost all date from the eighteenth and nineteenth centuries.[57]

The strongest hints of an early tradition of estate maps come from the hinterland of Venice, as seen in Lionello Puppi's *Andrea Palladio*, which reproduces property plans from the 1540s and 1560s.[58] As Denis Cosgrove has shown, the Venetian government was alert from the middle of the fifteenth century onward to the possibilities of maps in enforcing its control over Terraferma, its rich hinterland. But his work does not suggest that individual landowners were commissioning estate maps; rather, these were often ordered by the *magistratura sopra i beni inculti,* as part of the great sixteenth-century effort to drain and "improve" marshy areas.[59] Mapmakers like Cristoforo Sorte were entirely capable of meeting private commissions, but they seem almost always to have worked for one of the great Venetian public bodies. For instance, when in 1578 G. B. Remi mapped the Godi estates, it was at the request of the Council of Ten, in order to settle a lawsuit. It may be that future studies will reveal the existence of a Venetian landed patriciate, commissioning maps of their estates akin to the sixteenth-century English models, but such studies do not yet exist.

Perhaps in much of Italy (though not in Terraferma) landowning was relatively static, with the estates of the great valleys producing their traditional crops down the years primarily for local consumption; perhaps, too, the Italian agrarian economy was not well suited for the kinds of reform associated with the Agricultural Revolution in northern Europe. Be that as it may, present evidence suggests that few estate plans were drawn in Italy before the eighteenth century.

The Scandinavian Countries

The Scandinavian countries offer an interesting contrast with much of the rest of Europe.[60] There, seigneurial estate plans generally antedated the coming of state cadastral surveys, whereas in Scandinavia there were no local plans when the kings undertook their great seventeenth-century surveys. In Sweden, Gustav Adolph in 1628 set up the Lantmäteriet, which over the course of the century succeeded in mapping virtually all the productive areas of the country at a large scale. The equivalent in Denmark was the Matrikel of 1664 ordered by Frederick III (reigned 1648–70); this early survey had several successors. In Norway the corresponding Matrikkel was completed between 1665 and 1670, and served until the nineteenth century. Once these royal surveys had been made, local lords had little incentive to repeat work that was already well established.

Ireland

We know a good deal about large-scale topographical mapping in Ireland from the work of J. H. Andrews.[61] As in England "surveying without maps" was the rule until the later sixteenth century. At that time, topographical maps began to be made in connection with the various English incursions: the Munster plantations of the 1580s, the Ulster survey of 1609, and so forth. As time went by, these maps were produced at a larger scale, but they tended to include little internal detail; indeed, they often showed a ravaged countryside in which there was little internal detail to be shown. Even of the middle of the eighteenth century, Andrews writes: "For the student of roads and houses, in particular, the Irish estate map of the Gibson era will come as a disappointment. Such sparsely appointed interiors owe little to the estate cartography of other countries. In England, for instance, the best mapmakers offered a complete tableau of rural life: a village set among orchards, . . . [but it was not so in Ireland]" (155).

All this changed following the work of John Rocque, who in 1755–60 mapped the estates of Lord Kildare, setting new standards for detail that were carried on by his Dublin pupil Bernard Scalé and widely imitated for the next half century or so. Their maps were at a scale of about 1:5,000, or twice as large as had been normal until then, and they showed much detail. Down to the 1820s, their work was described as being "of the French school," and it does bear a close resemblance to the style of the French *arpenteurs,* shown in plate 7. Estate mapping in Ireland came to an end in the 1840s, when the great Ordnance Survey (1833–46) finally covered the whole island at a large scale and with great precision.

The incidence of surveyors in Ireland was quite high, for there were about 500 of them in 1750 and about 1,400 in 1840; more in the latter year, as Andrews notes, "than a time-and-motion expert would have been happy with" (244). They ranged widely in social class, with the poorer ones no doubt working primarily for farmers attempting to maintain their rights against landlords. Often, indeed, the surveyors measured small farms and fields, attaining an enviable degree of accuracy, for "it was by no means uncommon for pre–Ordnance Survey maps of Irish farms and townlands to come within 0.5 percent of the Ordnance Survey values" (235).

Conclusion

The incidence of estate plans in early modern Europe is quite varied, with many factors coming into play to explain their early presence or absence. By the late eighteenth century such plans were common in virtually all the countries that had some link with the general market economy. But it had taken a long time for the estate map to reach this level of diffusion.

Notes

1. Peter Barber in *Monarchs, Ministers, and Maps: The Emergence of Cartography as a Tool of Government in Early Modern Europe*, ed. David Buisseret (Chicago, 1992), 58.

2. See the remarkable (and so far unique) *Dictionary of Land Surveyors and Local Cartographers of Great Britain and Ireland, 1550–1850*, ed. Peter Eden (Folkestone, 1975–76).

3. A point demonstrated by Geoffrey Parker, "Maps and Ministers: The Spanish Habsburgs," in Buisseret, *Monarchs, Ministers, and Maps*, 124–52.

4. I should like here to acknowledge the help of the Casa de Velàzquez in Madrid, whose director allowed me to send a form letter to the roughly twenty French scholars working in Spain on subjects that might have revealed the existence of early estate maps. Almost all replied, almost completely in the negative.

5. On this theme see David Buisseret, contribution on Spanish peninsular mapping, in *The History of Cartography* (Madison, forthcoming).

6. For an example of the remarkable coverage of one region of the New World, see Barbara Mundy, "The Maps of the *Relaciones Geográficas* of New Spain, 1579–c. 1584," Ph.D. thesis, Yale University, 1993.

7. See Monique Pelletier, "De nouveaux plans de forêts à la Bibliothèque Nationale," *Revue de la Bibliothèque Nationale* 29 (1988): 56–62; Archives de France, *Espace français: Vision et aménagement, XVIe–XIXe siècles* (Paris, 1987), 16.

8. See Marc Bloch, "Les Plans parcellaires," *Annales d'Histoire Economique et Sociale* I (1929): 60–70, 390–98; Archives de France, *Espace français*, 15.

9. Quoted in *La Carte manuscrite et imprimée du XVIe au XIXe siècle*, ed. Frédéric Barbier (New York, 1983), 105.

10. These French maps have been studied by the abbé Roger Desreumaux; see, for instance, "Sources géographiques concernant la France conservées aux Archives Capitulaires de Tournai," *Horae Tornacenses* (1971): 275–91.

11. For a summary see Johannes Keuning, "Sixteenth-Century Cartography in the Netherlands," *Imago Mundi* 9 (1952): 35–63.

12. Roger Kain and Elizabeth Baigent, *The Cadastral Map in the Service of the State: A History of Property Mapping* (Chicago, 1992), 12–33.

13. See the bibliography under, for instance, Donkersloot–de Vrij, Fockema Andreae, Koeman, Vries, and Zandvliet.

14. See the exhibit catalog published by the Société Royale des Bibliophiles et Iconophiles de Belgique, *Les Albums de Croy* (Brussels, 1979).

15. Ibid. An "ingéniaire," or *ingénieur*, often made maps at this time, either for civil or (more often) for military purposes.

16. There is a good summary of these developments in Ernst Gagel, *Pfinzing: Der Kartograph der Reichsstadt Nürnberg, 1554–1599* (Hersbruck, 1957).

17. Globe production is well described in Oswald Muris and Gert Saarmann, *Der Globus im Wandel der Zeiten* (Stuttgart, 1961).

18. On these contributions see Robert W. Karrow, *Mapmakers of the Sixteenth Century and Their Maps* (Chicago, 1993); Peter Meurer, *Fontes cartographici orteliani* (Weinheim, 1991).

19. There is no thorough study, or even list, of these manuals. See the remarks of Fritz Hellwig, "Tyberiade und Augenschein: Zur forensischen Kartographie im 16. Jahrhundert," in *Festschrift für Bodo Borner* (forthcoming), 824.

20. This work is most accessible in the edition by Walter Strauss, *The Complete Drawings of Albrecht Dürer* (New York, 1977).

21. On Köbel's works see Josef Benzing, *Jacob Köbel zu Oppenheim, 1494–1533* (Wiesbaden, 1952).

22. On Pfinzing see Gagel, *Pfinzing;* Alfred Höhn, *Franken im Bild alter Karten* (Würzburg, 1986).

23. Pfinzing's counterpart in France, for instance, Philippe Danfrie, describes his new instrument the graphometer in *Declaration de l'usage du graphometre* (Paris, 1597).

24. In *The History of Topographical Maps: Symbols, Pictures, and Surveys* (London, 1980), P. D. A. Harvey calls these views "picture maps," and notes their popularity in sixteenth-century Germany (158, 168).

25. See the reproduction in *Niedersächsen in alten Karten*, ed. Heiko Leerhof (Neumünster, 1985), n. 7.

26. See Ernst Pitz, *Landeskulturtechnik, Markscheide, und Vermessungswesen im Herzogtum Braunschweig bis zum Ende des 18. Jahrhunderts* (Göttingen, 1967); Karl Brethauer, "Johannes Krabbe Mundensis," *Braunschweigisches Jahrbuch* 55 (1974): 72–89.

27. Pitz, *Landeskulturtechnik*, 147–150.

28. Reproduced in Leerhof, *Niedersächsen in alten Karten*, n. 4; see also *Der Deichatlas des Johann Conrad Musculus von 1625/26*, ed. Albrecht Eckhardt (Oldenburg, 1985).

29. Cited in Musculus, *Deichatlas*, 28.

30. Hans-Joachim Behr et al., eds., *Geschichte in Karten: Historishe Ansichten aus den Rheinland und Westfalen* (Düsseldorf, 1985).

31. See *Wilhelm Dilichs Landtafeln Hessischer Ämter zwischen Rhein und Weser*, ed. Edmund Stengel (Marburg, 1972).

32. See Fritz Bönisch, "The Geometrical Accuracy of Sixteenth- and Seventeenth-Century Topographical Surveys," *Imago Mundi* 21 (1967): 62–69.

33. Schweikher's manuscript was edited by Wolfgang Irtenkauf, *Der Atlas des Herzogtums Württemberg vom Jahre 1575* (Stuttgart, 1979).

34. Bönisch, "Geometrical Accuracy," 62–69.

35. See the article in Meurer, *Fontes cartographici orteliani*, 148.

36. See the detail in Ruthardt Oehme, *Die Geschichte der Kartographie des deutschen Südwestens* (Konstanz/Stuttgart, 1961), 41.

37. Ibid., 7.

38. See Ruthardt Oehme, *Johannes Oettinger, 1577–1633: Geograph, Kartograph, und Geodät* (Stuttgart, 1982).

39. See Oehme, *Geschichte; Wilhelm Schickard, 1592–1635: Astronom, Geograph, Orientalist, Erfinder der Rechenmaschine*, ed. Friedrich Seck (Tübingen, 1978).

40. Bönisch, "Geometrical Accuracy," 62–69.

41. See the endpapers of Seck, *Wilhelm Schickard.*

42. Some sheets from this magnificent map are superbly reproduced in *Cartographia Bavariae: Bayern im Bild der Karte*, ed. Hans Wolff (Weissenhorn, 1988).

43. For details see Gagel, *Pfinzing.*

44. See the two works by Hans Vollet, *Oberfranken im Bild alter Karten: Ausstellung des Staatsarchiv Bamberg* (Bamberg, 1983) and *Weltbild und Kartographie im Hochstift Bamberg* (Kulmbach, 1988).

45. On these developments in mathematics, geometry, and so forth, see Herbert Wunderlich, *Kursächsische Feldmesskunst, artilleristische Richtwerfahren, und Ballistik im 16. und 17. Jahrhundert* (Berlin, 1977).

46. Bönisch, "Geometrical Accuracy," 62–69.

47. See Ruthardt Oehme, "Johann Andreas Rauch and His Plan of Rickenbach," *Imago Mundi* 9 (1967): 104–7.

48. Bönisch, "Geometrical Accuracy," 62–69.

49. Oehme, "Rauch," 104–7.

50. I can say this with some certainty following the publication over the past twenty years of a large number of well-illustrated exhibit catalogs; see the bibliography under Aymans, Behr, Harms, Hellwig, Höhn, Jäger, Leerhof, Mertens, Reinartz, Vollet, Wolff, Zögner, and so forth.

51. See Eckhardt Jäger, *Prussia-Karten, 1542–1810* (Weissenhorn, 1982), 319.

52. Hennenberger's maps are described in Heinz Lingenberg, "Zur Geschichte der Kartographie Preussens," *Nord-Ost Archiv* 26–27 (1973): 3–13.

53. See "Administrative Mapping in the Italian States" by John Marino in Buisseret, *Monarchs, Ministers, and Maps*, 5–25.

54. Massimo Quaini, ed., *Carte e cartografi in Liguria* (Genova, 1986).

55. Sandra Faini and Luca Majoli, *La Romagna nella cartografia a stampa dal cinquecento all'ottocento* (Rimini, 1992).

56. M. A. Martullo Arpago, L. Castaldo Manfredonia, I. Principe, and V. Valerio, eds., *Fonti cartografiche nell'Archivo di Stato di Napoli* (Naples, 1987).

57. Walter Baricchi, ed., *Le mappe rurali del territorio di Reggio Emilia* (Casalecchio di Reno Bologna, 1985).

58. Lionello, *Andrea Palladio* ([Venice], n.d.), 259, 340. I owe this reference to the kindness of Andrew Morrogh.

59. Denis Cosgrove, *The Palladian Landscape: Geographical Change and Its Cultural Representations in Sixteenth-Century Italy* (Leicester, 1993).

60. For the first time, information on the Scandinavian countries is available in English in Kain and Baigent, *Cadastral Map*, 49–116.

61. Chiefly in J. H. Andrews, *Plantation Acres: An Historical Study of the Irish Land Surveyor and His Maps* (Belfast, 1985). Pages in text refer to this work.

Chapter TWO

ENGLISH ESTATE MAPS: THEIR EARLY HISTORY *and* THEIR USE *as* HISTORICAL EVIDENCE

P. D. A. HARVEY

Origins: The First Estate Maps

Estate map is here taken to mean a plan of landed property, drawn not for a particular occasion or for some closely defined purpose but for general reference. It was a work the estate owner could consult for detailed information about the lands it showed; or he might point to it with pride, seeing it as a graphic epitome of his property, wealth, and social position. Often it was clearly designed for display, beautifully colored and elaborately ornamented. Often signs of wear, and many added corrections and annotations, testify to long service as a functional tool of estate management.

Such estate maps appeared quite suddenly on the English scene.[1] There was no long course of development in which the genre gradually evolved, moving from the rudimentary to the elaborately detailed and sophisticated; those who made them did not feel their way toward plans of this kind. One of the most ambitious of all English estate maps before the nineteenth century, Ralph Agas's map of Toddington (Bedfordshire), dates from 1581, only a few years later than the earliest we know.[2] These date from the mid-1570s: the map of Queen Camel (Somerset) by Grindall Painter in 1573, the map of West Lexham (Norfolk) by Ralph Agas in 1575, and the map by Israel Amyce of the St. Paul's Cathedral estate at Belchamp St. Paul (Essex) in 1576 (figure 2.1).[3] Other early estate maps include the series produced by Ralph Treswell between 1580 and 1587 of lands of Sir Christopher Hatton in Northamptonshire.[4] Others, some perhaps earlier than 1575, will probably come to light. But it is clear that the estate map in England emerged fully fledged, almost certainly in the 1570s, and that from the start it was produced by various mapmakers; it was not the invention of a single innovator.

It did, however, have antecedents in the plans of the whole or parts of landed properties that were drawn for some particular, immediate purpose. These had a long history behind them; indeed, there are even a few medieval English examples, such as the thirteenth-century diagram of Wildmore Fen (Lincolnshire) that sets out the pasture rights agreed between Robert Marmion and Kirkstead Abbey.[5] After 1500, like all forms of local

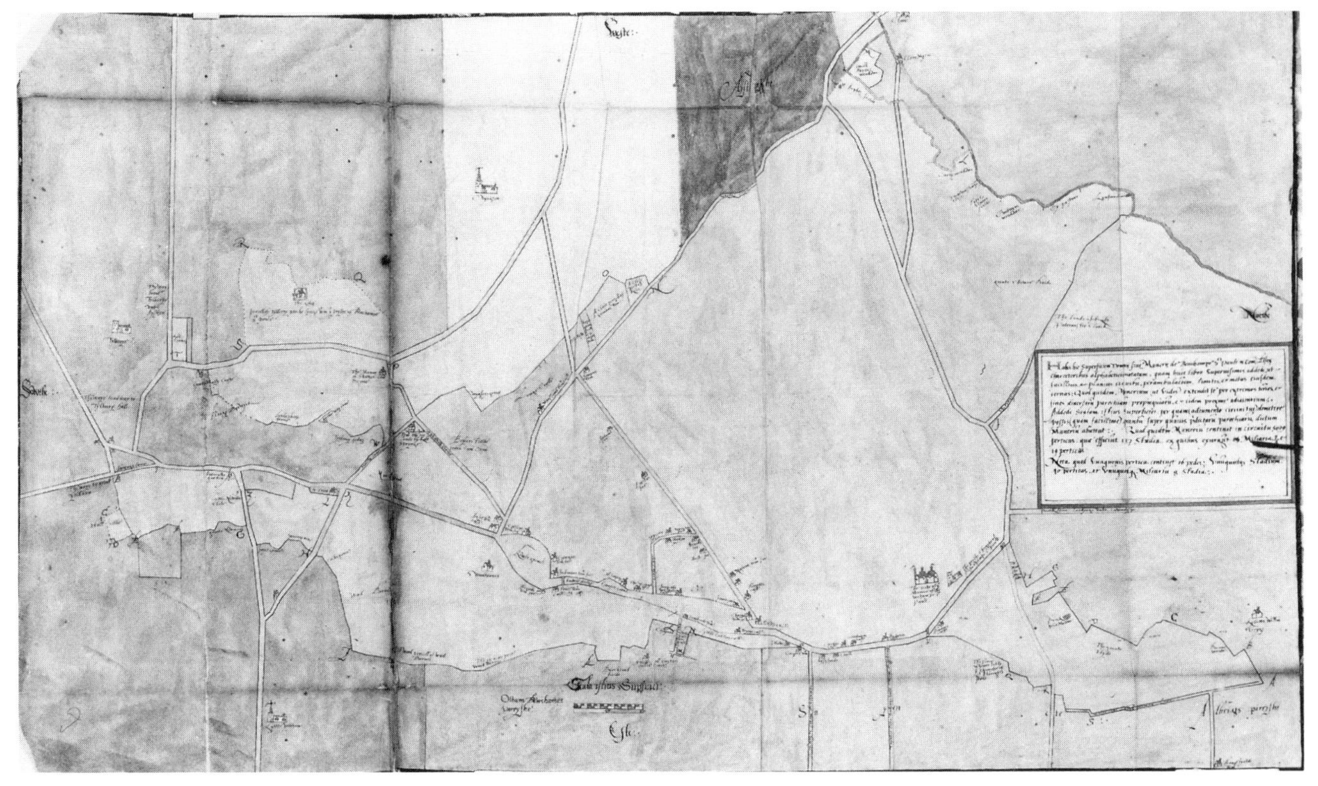

Fig. 2.1 Israel Amyce, St. Paul's Cathedral estate at Belchamp St. Paul (Essex), 1576 (Guildhall Library, London, MS 25517/1)

map, maps of landed properties became much more common and we have many examples, particularly from the mid–sixteenth century onward. Some are simply sketches. A plan of Sir Richard Edgcumbe's house and its surrounds at Cotehele (Cornwall) in the 1550s was drawn to show how access would be given to the local mills when a new deer park was created. It was a pen-and-ink drawing showing the river and the relevant roads and bridges, with rough outlines of church, houses, gates, and trees. Others include detailed information from written surveys.[6] Plans of parts of the fields of Dedham (Essex), drawn in 1573 perhaps in connection with a lawsuit, include a mass of notes giving the area and the tenant of each tiny plot of land and even page references to the survey that must have supplied the information.[7] Others again are carefully and elaborately drawn and colored. One of an area of marshland and its surrounds at Formby (Lancashire) was made probably for display in a court of law in 1557–58; the clumps of marsh flowers and the house, churches, mills, and woods are carefully executed but are all painted large and bold, to be seen from a distance.[8]

These earlier ad hoc plans of landed properties, made to meet some particular need, differ in another very significant way from the estate maps that first appeared in the 1570s. None is drawn to scale: each serves its intended purpose without any consistent relationship between distance on the map and distance on the ground, whether stated or implied. The Queen Camel map of 1573 was not drawn to scale, but other early estate maps are all scale maps. Certainly there had been earlier scale maps in England. The first date from the early 1540s: plans, mostly of projected fortifications, drawn by military engineers who probably brought the technique from Italy.[9] Over the following forty years quite a number of surviving scale plans of military works were drawn; there are particularly notable series from Berwick-upon-Tweed[10] and Portsmouth.[11] There was also a modest extension—no more—of the use of scale in two other sorts of map. One was a straightforward development of the plan of fortifications: military works would be shown in the context of what might be quite a large area of the surrounding countryside, and the whole map would be drawn to scale. Thus a superb scale map of the Isle of Sheppey (Kent), drawn about 1572 and signed with the unidentified monogram IM, was probably produced for military purposes and includes a minute plan of the fort at "Swale Nesse," but it is very much more than a simple plan of fortifications (figure 2.2).[12] The other scale maps of this period were of very much larger tracts of country, 200 square miles or more, drawn for a variety of purposes: for military planning, like one of part of the Anglo-Scottish border in 1552; for use in a lawsuit, like one of Ashdown Forest (Sussex) about 1563; or for general reference, like one of County Durham in 1569–70.[13] The value of consistent scale was appreciated sooner in these maps of relatively large areas than in maps and plans of more limited scope, apart from those drawn by military engineers.

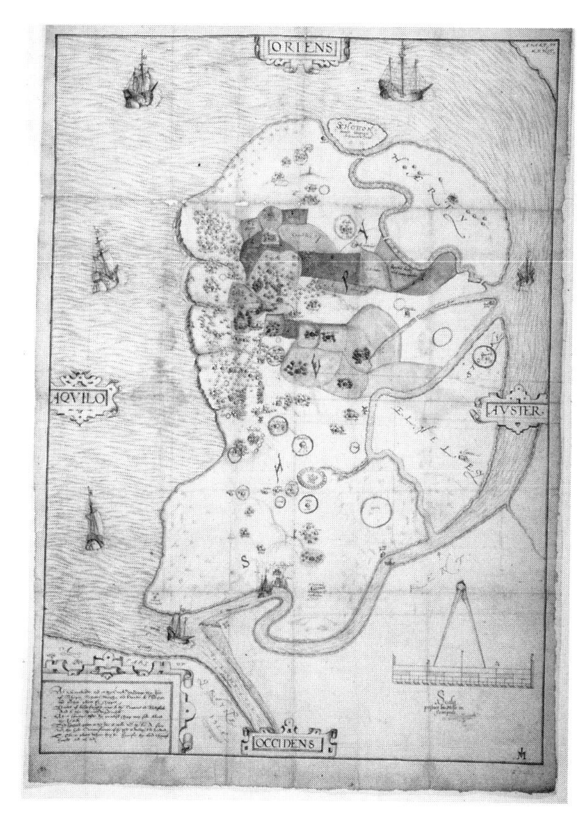

Fig. 2.2 IM, Isle of Sheppey (Kent), 1572 (Public Record Office, London, MPF 240)

The printed county maps of Christopher Saxton fall neatly into this pattern. They are scale maps of substantial areas: every one has a scale bar. The individual maps bear dates from 1574 to 1578, and the complete collection was published as an atlas in 1579.[14] The number of surviving copies shows that the maps sold well, and they must have done much to familiarize Englishmen with the value of maps in general and with the concept of scale in particular. It is not impossible that they played a large part in the introduction of estate maps; that is to say, landowners commissioned maps of their properties because they had bought or seen Saxton's maps of their counties. The coincidence of dates is interesting and may be significant.

Whatever its cause, we should clearly look to the landowners as well as to the surveyors for the impetus that produced the sudden emergence of estate maps in England. Ralph Agas wrote in 1596 how the value of a map in setting out a surveyed boundary had suddenly occurred to him some twenty years earlier;[15] but from what we know of the men who produced the first estate maps—their education, their interests, their writings—we would have expected them to have been long aware of maps, aware even of the possibility of mapping to scale. There was no development in techniques or instruments that made it more practicable to draw estate maps in the mid-1570s than it had been fifteen or twenty years before. All that was new was the idea, the application of mapping to estate management; and what happened in the 1570s was probably that the surveyors who were to produce the maps succeeded in convincing the landowners who were to pay for them that this idea was a good one.

We should remember that every estate map had two parents: the commissioning landowner who fathered it and the mapmaking surveyor who brought it to birth. We can only guess the part that each played in the origin of estate maps in England, but in following their development down to the early eighteenth century, we can distinguish fairly clearly the role of each parent.

Development: The Role of the Surveyor

By the 1570s estate surveyors in England had a long history behind them. Fully measured surveys of estates first appear in the early thirteenth century, and by at least the early fourteenth we find some men who, while far from being professional surveyors, were apparently considered to have special knowledge or skill in measuring and describing lands. The products of their work were written surveys, formal descriptions of estates that the owners would use in the work of management, precisely defining their property and revealing its potential, providing a yardstick with which actual production and profit might be compared. These written surveys might be long and elaborate, giving details of sometimes thousands of individual plots of arable, besides meadow, pasture, and woodland. Mapping played no part in medieval surveying; it is just possible, no more, that fifteenth-century surveyors occasionally sketched rough maps as an adjunct to their working notes, but a map never formed part of the final survey. The work of the estate surveyor was to measure and describe lands, not to draw maps.[16]

This was the case until the 1570s. From the beginning of the sixteenth century the understanding and use of maps spread fast, and we have seen that sketch maps of pieces of landed property might be made for some particular purpose. No doubt estate surveyors made some maps of this kind, just as they probably made use of sketch maps in drafting their surveys. But the work of a surveyor was still to measure and describe lands in written surveys: he was no more a mapmaker than the landowner who employed him. We see this in published treatises on surveying. The earliest of these, John Fitzherbert's *Boke of Surveyeng* (1523) and Richard Benese's *Maner of Measurynge* (1537), like their medieval predecessors that circulated in manuscript, make no mention of drawing maps: they are concerned with measurement, particularly how to measure areas of variously shaped plots. The same is true of mid-sixteenth-century treatises on measuring distances and areas, by Leonard and Thomas Digges in 1556 and 1571, by Valentine Leigh in 1577, and by William Bourne in 1578. Those that mention mapmaking—and those of

1571 and 1578 refer to making scale maps—see this as part of military science, not as anything to do with the estate surveyor. The earliest book that envisages mapmaking as part of a surveyor's work was published in 1582: *A Discoverie of Sundrie Errours and Faults Daily Committed by Landemeaters, Ignorant of Arithmetike and Geometrie* by Edward Worsop.

It was thus as measurers and describers of lands, not as mapmakers, that surveyors began in the mid–sixteenth century to be seen as specialists having particular skills and expertise. This reflects the development of new techniques and new instruments for mensuration, and from then on many surveyors were highly qualified and experienced professionals. But much work was still done by part-time, even amateur, practitioners, and in the eighteenth century some treatises were still recommending methods of measuring that were far from adequate.

By the end of the sixteenth century even the most professional still saw mapmaking as only a part, and by no means the most important part, of an estate surveyor's work. The most map-minded might see it as an essential part. Agas wrote: "No man may arrogate to himselfe the name and title of a perfect and absolute Surveior of Castles, Manners, Lands, and Tenements, unlesse he be able in true forme, measure, quantitie, and proportion, to plat [i.e., draw maps of] the same in their particulars *ad infinitum*."[17] But the very fact that Agas had to assert this suggests that there were those who thought otherwise. Even so prolific and proficient a mapper of estates as John Norden made it clear in his *Surveyors Dialogue* (1607) that mapmaking was not "the chiefe part of a Surveyors skill": other abilities were "more necessary." Besides actually measuring and describing land he had to interrogate local inhabitants about bounds and rights, he had to be able to examine records, including those in Latin and in antique handwriting, he had to value lands before they were leased or sold, he had to advise the landowner how to run his estate and how to get the most profit from his property.[18] In Norden's eyes the surveyor was a highly skilled professional, entitled to much respect from estate owners and others, but not simply because he ought to be a competent cartographer. It should be added that Norden wrote as a propagandist: he probably idealized the estate surveyor's skills and exaggerated his importance and status.

Since estate maps were made by surveyors, whose work was based on the measurement of land, it is easy to see why, right from the start, they were always drawn to scale. To the landowner, once he had mastered the idea, the scale map offered much more than a map drawn without consistent scale. And for the mapmaker there would be no difficulty: he would already have by him all the information needed for a scale map, which would simply be a graphic representation of what he had had to do anyway. Some estate maps include pictures of the surveyor at work in the field, underlining their origin in the careful measurement of land. An early and particularly interesting example is on a map of 1659 from Bryanston (Dorset) (figure 2.3). Here, as so often on early estate maps, the outline plans of the fields are enlivened by typical scenes: herds grazing, men at work plowing, sowing, and harvesting, juxtaposed in cheerful disregard of the due order of seasons. At one edge of the map is another scene, the surveyor's team in action, half a dozen strong, with their names. "Mr Stratton," the surveyor, stands with his stick, supervising the operation, "Mr Fisher" is holding a pole, "Mr Andrews" is carrying what is probably a plane table, "John Heath" holds the end of a chain, and so on.[19] Measurement and estate mapping were soon seen as parts of a single process, so that certainly by the

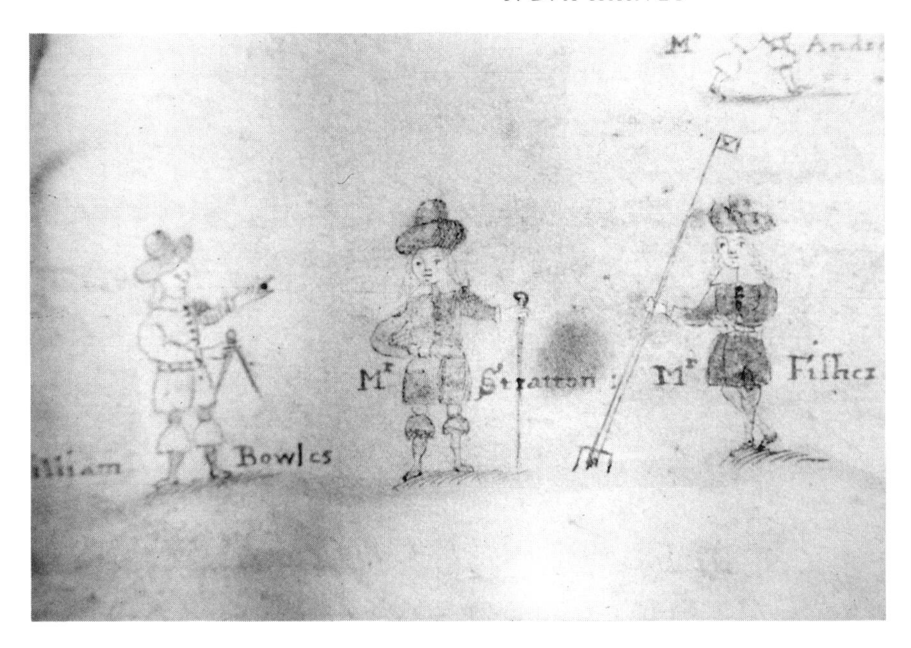

Fig. 2.3 N. Stratton, Bryanston (Dorset), 1659 (detail) (Dorset Record Office, Photocopy 500)

late seventeenth century the estate surveyor would commonly be thought of as a mapmaker. The connection that we take for granted between surveying and mapping was already established.

This appears clearly in two developments that occurred early in the history of estate mapping in England. One was that surveyors began to be brought in to make scale maps in circumstances that earlier had called for nothing more than a simple sketch map, drawn without measurement to no consistent scale, a map that did not need the special skills of the surveyor. We see this in maps made in connection with lawsuits. In the records of the duchy of Lancaster is a notable series of sixteenth-century maps that were made for production in the duchy's own courts—often on the orders of the court itself—in suits concerning its tenants. The earliest recognizably drawn to scale are a group of three dating from 1580–81, of lands in Accrington, Penwortham (both Lancashire), and Barnoldswick (West Riding of Yorkshire). They all have scale bars but are so crudely drawn and painted that we might suppose this to be no more than a gesture toward the idea of the surveyed map.[20] However, a document attached to the Accrington map implies that, although the whole map covers about 1,000 acres, only the particular pieces of land involved in the case, some 50 acres in all, were measured and drawn to scale; the rest must have been sketched in to set these lands in their local context. Of the lands at issue "the moste parte" took some four or five hours to survey on 29 May 1581, and the finished map was returned to the court the following day. Probably the surveying was fairly rudimentary.[21] All the same, it is very interesting that, just as estate maps were coming rapidly into fashion, measurement on the ground and consistent scale began to be demanded in this other sort of map as well.

An interesting group of maps of Woodmancott Farm near the Candovers (Hampshire), property of Winchester College, neatly illustrates the course of this development. Only one is dated, but they all belong to the late sixteenth century, and their chronological order probably corresponds to their undoubted typological order. First are two similar but independent plans of the farm, not drawn to scale, which the catalog of the college archives suggests were drawn in connection with lawsuits (figure 2.4). Then comes a general estate map of "the manor of Candever," a much wider area including the farm; it was drawn to scale, with a scale bar, and is signed by Ralph Treswell in 1588 (figure 2.5).[22] Finally, there are two copies of a map of the farm alone, probably again for a lawsuit, but this time drawn to scale (figure 2.6); with them are a set of six of the surveyor's drafts from which the map was constructed.[23] The estate map, drawn to scale from the surveyor's measurements, set a new standard in mapping of properties, to which other sorts of map now began to conform.

The second development was that, once estate surveyors had successfully introduced the map as a way of setting out the information from a general survey, giving it graphic realization, drawing maps came more and more to be seen as a normal part of their work. Thus they would often produce scale maps when they were employed on other lesser

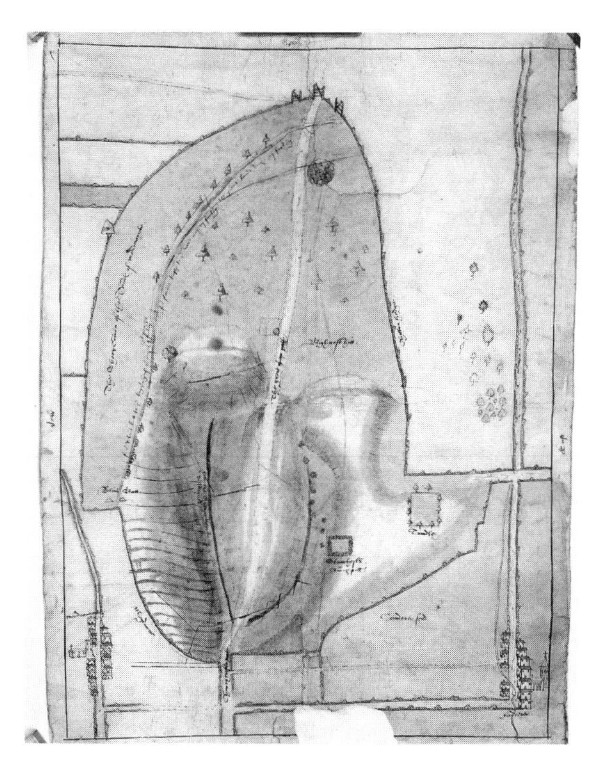

Fig. 2.4 Anonymous, Woodmancott Farm (Hampshire), late sixteenth century (By permission of the Warden and Scholars of Winchester College, muniments, 21446)

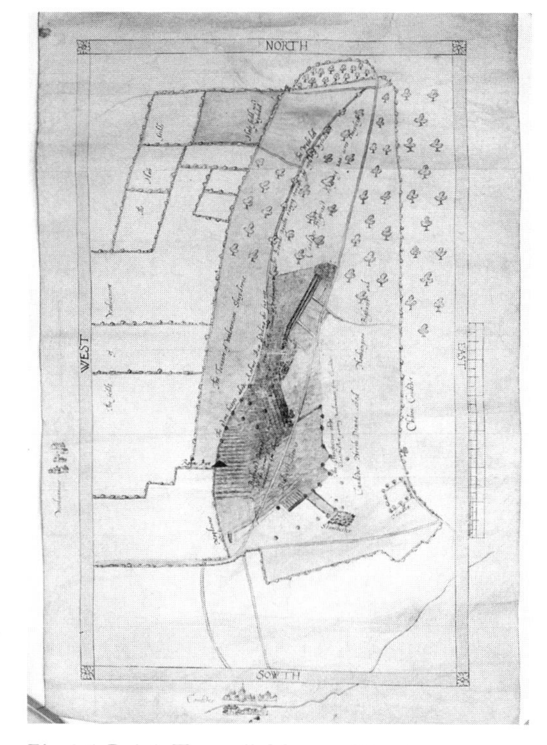

Fig. 2.5 Ralph Treswell, Manor of Candover (Hampshire), 1588 (By permission of the Warden and Scholars of Winchester College, muniments, 21443)

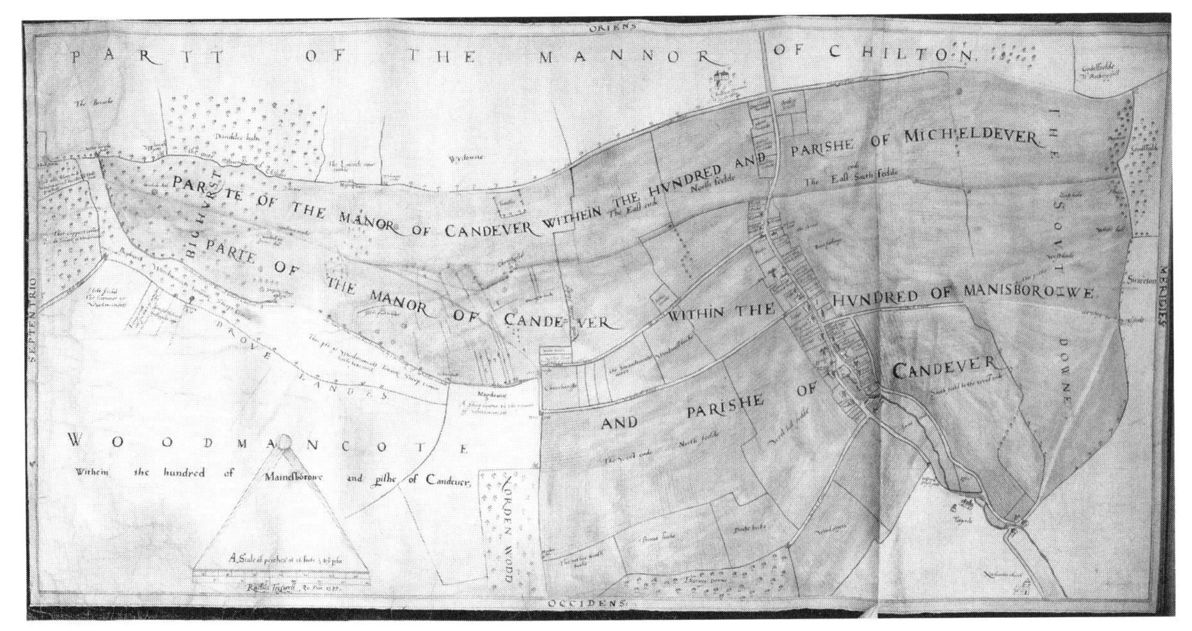

Fig. 2.6 Anonymous, Woodmancott Farm (Hampshire), late sixteenth century (By permission of the Warden and Scholars of Winchester College, muniments, 21444)

jobs for the proprietors of landed estates: measuring a piece of land that was to be sold or divided, for instance, or calculating the effects of some projected improvement. Probably this came about quite quickly, but the chronology has yet to be investigated. Often it is not easy to tell why a particular map was made: the coloring or other conventions may not be explained, and where a map covers a considerable area of property, we can sometimes only guess that it was produced for some ad hoc purpose and not as an estate map for general reference.

This uncertainty can be illustrated by some maps from south Warwickshire at the end of the seventeenth century and the beginning of the eighteenth. One by John Garfield about 1700 shows the river at Stratford-upon-Avon with a substantial area on either side, including meadows, an orchard, the parish church, and houses (figure 2.7). Some lands are outlined in various colors, but their significance is not explained and there is no title, though one seems to have been erased from a panel in one corner. We can only say that it is likely, but not certain, that this map was drawn for some particular occasion of which we know nothing. We can be more certain about four plans of lands at Whitchurch and Crimscote drawn by the same John Garfield, one in 1687, one probably about the same time, and the other two in 1709 and 1712. They are functional outline plans, each showing only one or a small group of plots, covering from 8 to 20 acres in all, with notes of their areas and just a few other details. Why each was drawn we again do not know, but here there can be little doubt that it was for some immediate purpose.[24] On the other hand, a plan by James Fish of the churchyard at Snitterfield in 1697

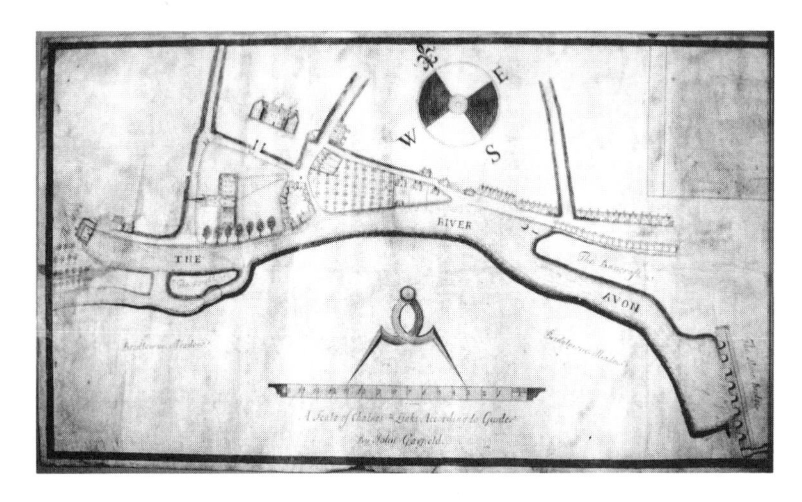

Fig. 2.7 John Garfield, the river at Stratford-upon-Avon (Warwickshire), c. 1700 (Warwickshire County Record Office, CR 85, bundle 4)

has a key that makes it quite clear why it was drawn: it was to show a proposed extension to the churchyard (figure 2.8). Probably all these maps served some equally clear and immediate purpose, now hidden from us, reflecting the way that surveyor and landowner might now use scale maps in the everyday work of running an estate.

The surveyor might have been a part-time occasional practitioner or a full-time professional. In either case he normally worked in a fairly limited area, and we find that in any generation the estate mapping within a single county is dominated by a very few names. Surveyors were already established in this pattern of work by the late sixteenth century, so we find it in the earliest estate maps. The work of Christopher Saxton provides a notable early example. Most of his fifty-odd known estate maps and surveys are of places in Yorkshire, but about a dozen are from other parts of eastern or southeastern England, one is from Shropshire, and one from Lancashire. The seventeen maps or surveys that his son Robert is known to have produced are all from Yorkshire.[25] Again, Thomas Kington and George Withers are two names that recur in the estate maps from early-seventeenth-century

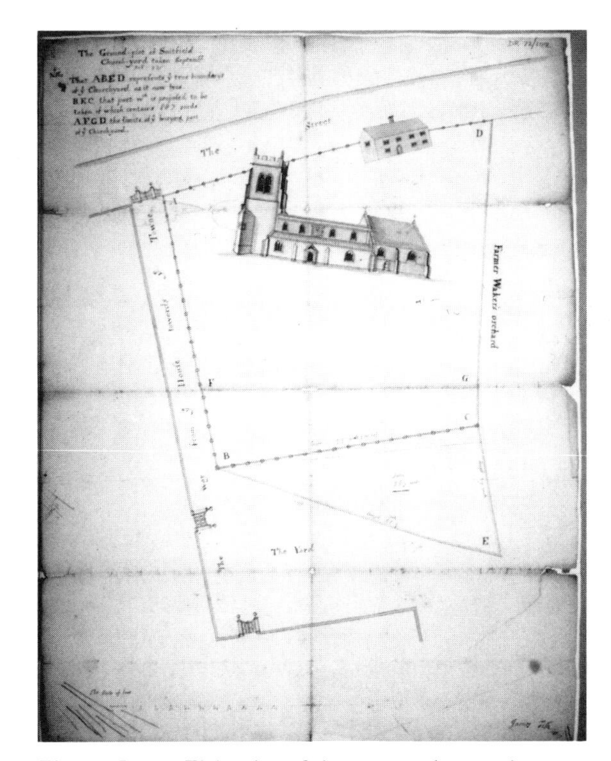

Fig. 2.8 James Fish, plan of the proposed extension to the churchyard at Snitterfield (Warwickshire), 1697 (Warwickshire County Record Office, DR 72/102)

Hampshire. But none of these achieved the local dominance that we find in the work of James Fish in Warwickshire over a long period, from the 1680s to the 1730s; some thirty of his maps survive from all parts of the county. In addition we have a few maps that he made in other counties, mostly adjoining Warwickshire but including Somerset (one map of Sutton Bingham, near Yeovil). This pattern was probably typical of the well-established professional surveyor at this time: most work would come from what might be quite a small number of landowners in his own area, but either because they owned some distant properties or because personal contact or reputation brought commissions elsewhere, some of his work would be done far from home. Whereas Fish worked all over Warwickshire—throughout his career he was based at Warwick—two of his contemporaries operated in smaller areas. John Garfield worked, as we have seen, in the south of the county, and Robert Hewitt in the north; each produced some eight or nine surviving maps.[26]

In the course of the seventeenth century, as making maps became more and more a normal, accepted part of a surveyor's work, there developed a general proficiency, an increasingly professional approach to producing them. This appears in various ways. One is greater uniformity of style. The earliest estate maps differed a lot one from another. The overall impression of the maps of Ralph Agas, for example, with their careful finicky lines, little use of color washes, and lettering in red or green, is quite different from the contemporary maps of John Norden, probably just as carefully measured and drawn but with bold overall coloring and with lettering in black ink. It would be wrong to suggest that later surveyors' maps all looked alike: there were many individual quirks of convention and style. But by the end of the seventeenth century the differences were less extreme than they had been a century earlier.

One interesting instance of the professionalism of surveyors who were producing maps regularly is the use of printing blocks for conventional signs and decorative features on the maps they drew. An early and rather crude example is a map of Thankful Frewen's estate at Northiam (Sussex), drawn by Giles Burton in 1635: the circle around the number given to each field, the scrolls containing their areas, the houses, even horses and cattle and a man carrying a staff, have all been stamped on the map from engraved woodblocks.[27] We find the same technique on a map of Sir John Mill's manor of Langley in Eling (Hampshire), drawn by Nicholas Ayling in 1692, but here the printing is confined to the decorative elements (figures 2.9, 2.10). The elaborate floral border is printed, using probably seven blocks, as are the ships on the water and the thirteen figures of deer in the New Forest; nine different blocks were used, showing deer walking, browsing, and so on, which suggests that Ayling was accustomed to having landowners in the forest as his clients. The blocks are finely engraved, and (unlike Burton's map) it is not at once obvious that these features were printed. On the other hand, the houses and other nondecorative features on Ayling's map are all drawn by hand.

Yet however skilled surveyors became as cartographers, however much drawing maps was seen as a normal part of their business, still in the early eighteenth century, just as a 150 years earlier, the primary work of a surveyor was to measure and report on land. He might present the results of this work in a map or in a written report; how it was done, which method was chosen, was between him and his client. In fact, throughout this period there is great variety in the roles of the written survey and the estate map. All the information gathered

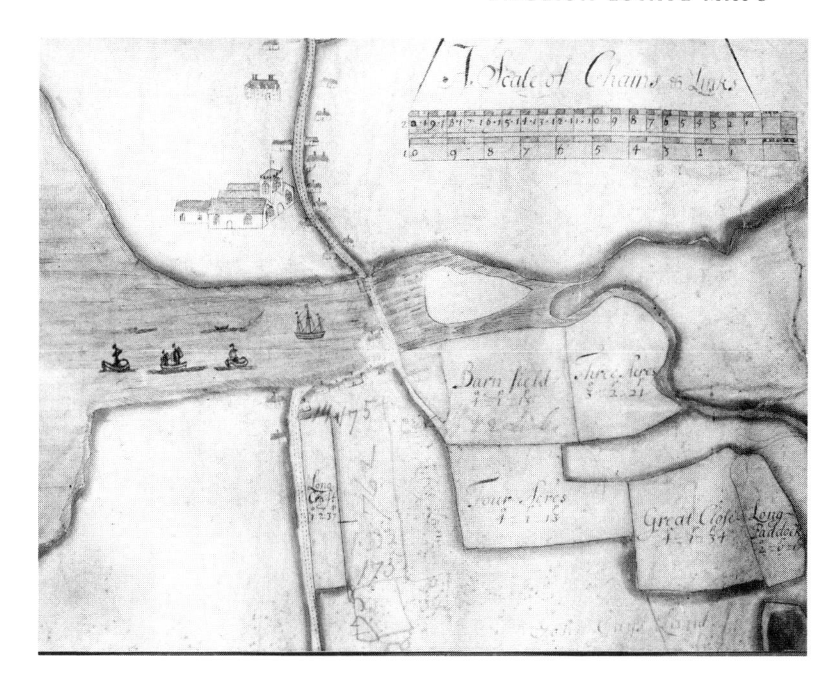

Fig. 2.9 Nicholas Ayling, detail from a map of the manor of Langley (Hampshire), 1692 (By permission of the Warden and Scholars of Winchester College, muniments, 21335)

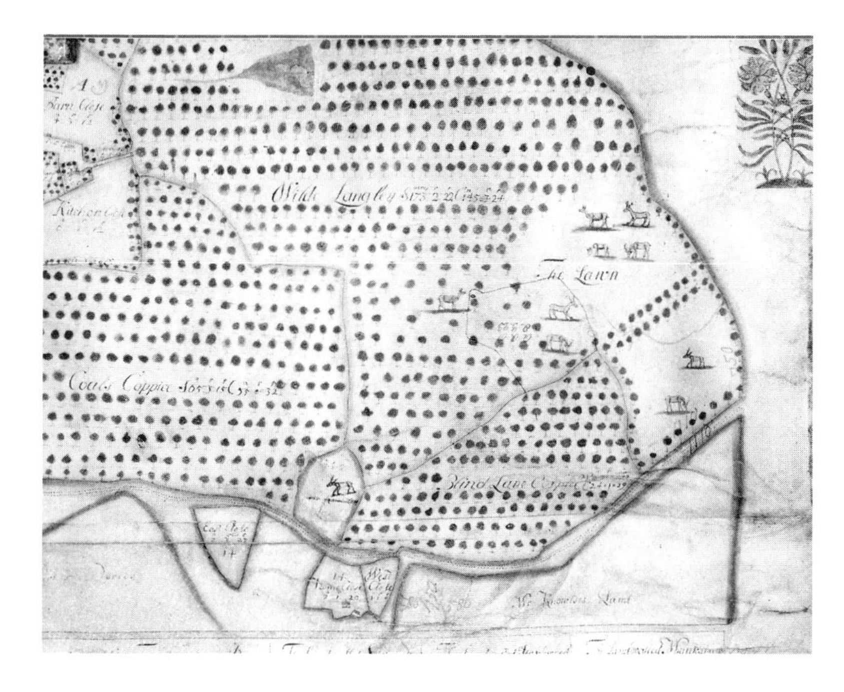

Fig. 2.10 Nicholas Ayling, detail from a map of the manor of Langley

might be entered on the map, which would then entirely replace the written survey. This seems to have been the case in Ralph Agas's 1581 map of Toddington (Bedfordshire), drawn on twenty pieces of parchment stuck together to form a single sheet measuring 8.3 by 11.5 feet, which was then attached to a roller; at the edge of the map are lists and summaries that supplement the notes of area and tenure written on each plot of land, and there was probably no other product of the survey—all the information is contained on the map itself.[28] The same is true—to take another early example—of the anonymous map of Bolton Percy (West Riding of Yorkshire) of about 1595. This is on thirty-eight sheets bound as a volume, but here detailed information about the areas and tenants of the lands is given in lists beside the maps, not on the map surface.[29]

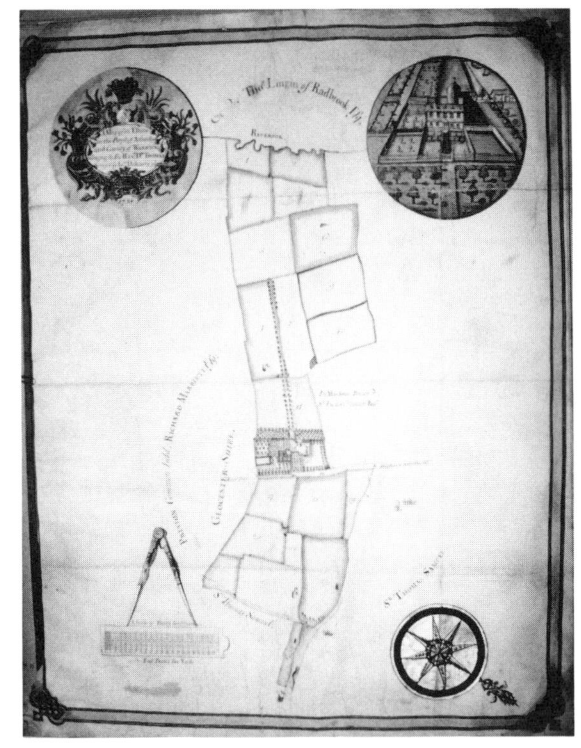

Fig. 2.11 John Doharty, Dr. Thomas's lands at Atherstone-on-Stour (Warwickshire), 1738 (Warwickshire County Record Office, CR 1945)

Often, however, both a written survey and a map or maps were produced, and the relationship between them could differ greatly. The written survey might simply explain or elaborate on the information given on the map. We see this in two very different examples from Warwickshire. James Fish's magnificent map of the earl of Rochester's estate at Kenilworth in 1692, some 6 feet square, is accompanied by a forty-four-page survey book that summarizes and tabulates information already given on the map but also adds the rents due and copies of relevant medieval title deeds.[30] The small but handsome map of Dr. Thomas's lands at Atherstone-on-Stour that John Doharty drew in 1738 has on a separate piece of parchment what amounts to a key to the map: it lists the names, areas, and use of the plots of land that on the map are simply given numbers (figure 2.11).[31] On the other hand, the map might do no more than illustrate what was primarily a written survey. A written description of first the demesne and then the tenants' lands of the manor of Bisterne (Hampshire) in 1591 occupies twenty-seven leaves of a bound volume and is followed by a single opening on which are drawn nine plans, in varying states of completion, of particular parts of the estate.[32] From Ansty (Warwickshire) we have a survey written in 1599 and a map that may well have been drawn up at the same time by way of illustration but that gives none of the information about the areas or tenure of plots that is given in the survey.[33] Also from Ansty is a book of some ten leaves setting out the precise measurements and other details of the enclosure and redistribution of lands from common fields in 1602, the work of Richard Bankes, "Practitioner in the Matthamatticks"; no map accompanies it.[34] Of course, where a map and a written survey were drawn up together but—as so often—on separate sheets or books, there is always the risk that the two might become divorced over the years and one of them lost from sight or destroyed; maps often have a different archival history from the other records of an estate.[35] We might then wrongly suppose that the survivor of the pair, whether map or written survey, was the sole product of the surveyor's work.

However, written surveys without maps were being produced throughout the period. One undoubted example, made in 1721, is of the lands of Richard Mariett and his tenants in Preston-on-Stour (Warwickshire), a roll of six pieces of parchment, 11 feet long, describing in turn over three thousand strips of common-field arable or other land (figure 2.12); even in the early eighteenth century the estate surveyor would not be thinking of a map as the only way to present his work. On 4 May 1685 James Fish wrote from Warwick to Davenport Lucy, then staying in Covent Garden in London, about work he had been doing on Lucy's lands at Hampton Lucy and Charlecote (both Warwickshire):

> The last weeke I was preingag'd in business that I could not defer, without prejudice to the Gent': by whome I was imploy'd, and having no more time then this day to peruse any thing of Hampton Lucy or to cast up that part of Charlcote that I have allready survey'd I humbly beg yor: pardon for the imprefect Account I have here inserted viz:

	A	R	P
Charlcote Field is about	500:	00:	00
with 5 little Mead: as apart thereof containing about	015:	00:	00

> There's about 60 Acres of y^{e} field next Newbold that is very course Ground, the rest by experienc'd and impartiall men is valu'd a^{t} as

Fig. 2.12 Written survey of the lands of Richard Mariett and his tenants in Preston-on-Stour (Gloucestershire), 1721 (Warwickshire County Record Office, CR 936, box 15)

> good a rate as any common field land thereabouts and lies so accommodated with all Conveniences that makes it capable of good improvement by inclosing.

Continuing, he mentions that he has made a "collection" of the tenants and their tenures and will transcribe it if wished, also a rental of the estate and an inventory. He concludes: "The Totall of y[r] whole Estate viz: Hatton Fullbrooke Hampton Lucy Charlcott & Hunscote is about 4500 Acres In my next I shold give a more particular account and insert those Remarques I have made from y[e] opinion of experienced men tho they all owne themselves farr short of yo[r] great judgment in these affaires." In all this there is not a word about drawing a map.[36]

On other occasions Fish did draw maps of parts of the Lucy family's property; they knew perfectly well the value and use of maps in estate management, but it was not only in order to draw maps that one employed a surveyor. There was other work for him to do as well. Some other landowners probably never saw any need for the trouble and expense of having maps made of their estates. After all, as John Norden made the tenant farmer say in his *Surveyors Dialogue* of 1607, "is not the Field it selfe a goodly Map for the Lord to looke upon, better than a painted paper? And what is he the better to see it laid out in colours?" (15). Some landowners would have agreed with this point of view. Others did not. As we have seen, every estate map had two parents: the surveyor who gave it to the world and the landowner who was responsible for its conception. The landowner's role in the early development of the estate map in England was just as important as the role of the surveyor.

Development: The Role of the Landowner

The number of early estate maps differs greatly from one county to another. For Warwickshire in the Midlands a full census of manuscript maps has been made, and another is in progress for Hampshire in the south.[37] From Warwickshire there are forty estate maps earlier than 1701 in repositories accessible to the public, from Hampshire only sixteen; and from County Durham in the north of England there is only one.[38] These three counties are of different sizes, and if we adjust the figures to give the number of maps per 1,000 square miles the result is Durham 1, Hampshire 11, Warwickshire 42. It would be wrong to attach importance to the precise figures. In some cases there is room for doubt whether a map is or is not an estate map by our definition. The total numbers are small, and the simple accident of a map's having a better chance of survival on one estate than on another may affect the figures. And, of course, reckoning only those in accessible repositories may significantly distort the picture. All the same, the figures for these three counties are so very different that they must point to real differences in the number of early estate maps that were produced.

Much more work will have to be done in individual counties before we can fully appreciate, let alone explain, these differences—careful, detailed work of the sort that Sarah Bendall presents here on the maps from Cambridgeshire. But some preliminary comments can be made on the evidence from Durham, Hampshire, and Warwickshire. First, it is unlikely that the lack of maps from Durham points to a contrast between north and south, that understanding and use of estate maps spread late to the north of England. On the contrary, in neighboring Yorkshire, estate maps appeared early on the scene; I have already referred to the work of Christopher Saxton and to the map of 1595 from Bolton Percy. What is significant, however, is that most of County Durham was owned by a very small number of large landowners, and the archives of only two of them are in publicly accessible repositories. Warwickshire had a larger number of substantial landed estates, but if we had no maps from three of them, the Warwick Castle estates and those of the Leigh and Newdegate families, our figure of surviving and accessible maps from the county would be very different. Even allowing for the accident of loss or survival of particular estate archives and maps, it seems clear that probably the majority of landowners did not have maps made of their properties. We see this in the early estate maps made for Oxford colleges. Some, notably All Souls and Corpus Christi, commissioned considerable series of maps; others seem to have had maps made only as occasion demanded or not at all.[39]

The two major County Durham landowners whose archives are publicly accessible contain not a single early estate map, and it is interesting that both these landowners are ecclesiastical: the dean and chapter of Durham and the bishop. One of the earliest of all English estate maps, of Belchamp St. Paul (Essex), was made for the dean and chapter of St. Paul's Cathedral in 1576 (figure 2.1). All the same, a first impression suggests that the church was less inclined than other landowners to commission estate maps. Glebe terriers—detailed accounts of the lands held for the incumbent's support—which the parish clergy were required from time to time to submit to the archdeacon or bishop, practically always took the form of written surveys. Of some 120 Warwickshire glebe terriers dating from 1585 to 1675 not one is even accompanied by a map;[40] from County Durham one, for Stanhope in 1663, was submitted as a map, without a written description, and from Hampshire too there is just one map among the glebe terriers before 1750, a sketch plan of 1718 showing glebe land of a church within the city of Winchester.[41] The force of tradition and the difficulty of mapping small properties made up of scattered plots and strips must have contributed to this; but behind it may lie a particular reluctance on the part of church administrators to use maps in estate management. The point needs investigating.

So too does the relationship between estate map, written survey, and manorial court. From the sixteenth century onward we have records of manorial courts meeting in what was called a court of survey. This was held to compile a survey that listed the manorial tenants, what they held and by what tenure, and what they owed in annual or other payments. If we are to believe the treatises on holding manorial courts, which were still being published in the nineteenth century, neither mapping nor mensuration played any part in this operation: the information was gathered from the evidence of the tenants themselves and from deeds and other documents.[42] However, John Norden in 1607 envisages the court of survey as being summoned and held by the professional surveyor whom the landowner has brought in to survey the estate; in it he interrogates the tenants and examines the documents they produce—but then he goes on to measure and map the manor as part of the same process.[43] Some early estate maps and measured surveys were certainly drawn up as business of the manorial court. The survey already mentioned of Bisterne in 1591, with maps attached, was drawn up in a court of survey; so was the survey of Ansty in 1599, which may or may not have had an accompanying map; so too was Ralph Agas's map of Toddington in 1581, which may well have stood on its own without a written survey. However, most English estate maps have no apparent connection with a manorial court. Throughout the seventeenth and eighteenth centuries manorial courts themselves were in decline: their effective business diminished, proceedings became formalized, and many courts simply stopped being held. But even if the landowner held a court for the lands that he wished to have surveyed or mapped, it may well have seemed an unnecessary complication to involve it in the business. How far estate mapping and manorial courts went hand in hand, whether it was the decline of manorial courts that separated them, and whether they had ever really been closely connected are questions that future research may well answer.

In any case the owner of a mapped estate need not have been the lord of a manor. We have maps of estates of every sort, from those of great lords down to those of quite humble tenants. On the one hand we have John Norden's survey in 1607 of the King's Honour of Windsor (Berkshire), an area measuring some 16 miles by 20: a parchment book, handsomely bound and with the royal arms at the front, containing twenty colored maps with notes that include a table of the number of deer on the estate (figure 2.13). At the other extreme is a little book of six paper leaves containing plans by John Tipper of the 20 acres owned by Robert Smith of Coundon (Warwickshire) in 1694; on the first page is a pleasing pen-and-wash picture of his modest timber-framed house with attached stable (figure 2.14). Whether the estate was large or small, we still do not know where the idea for mapping it came from, whether it was the landowner who sought out

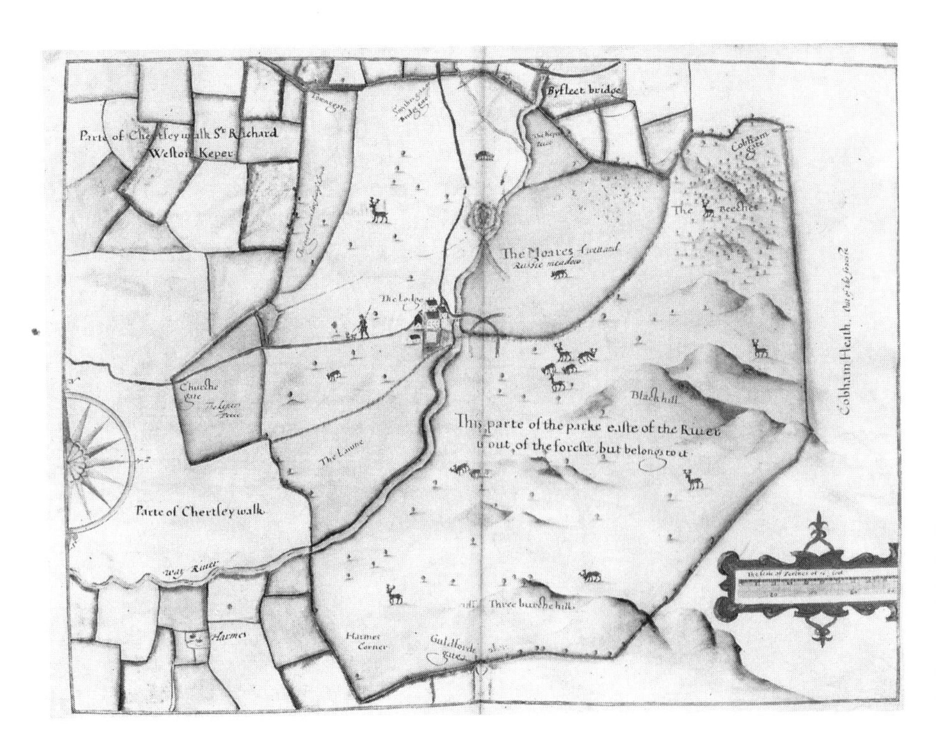

Fig. 2.13 John Norden, detail from a survey of the King's Honour of Windsor (Berkshire), 1607 (British Library, Harley MS 3749, fol. 16)

Fig. 2.14 John Tipper, page from a book of plans showing land owned by Robert Smith of Coundon (Warwickshire), 1694 (Warwickshire County Record Office, CR 299/574)

the surveyor for the purpose or the surveyor who took the initiative in selling his services to a wide range of customers.

For the landowner there were clear practical advantages in having a map of his properties. Some were described by Ralph Agas in *A Preparative to Platting of Landes and Tenements for Surveigh,* a twenty-page pamphlet published in 1596. One was that "the bounders and abbuttals of every particular is so wholy put downe, as no booke may bee comparable with the same"; another that "[i]f you will sever any fielde or cloase into two or more parcels: the Scale will readily bewray how many perches, & feet shall performe the same, and where may be the rediest cut"; and another that it eliminated the risk of losing lands because change of name or of use meant that they could not be identified in old written records (14–15). Certainly many landowners saw these and other advantages and made use of maps in the practical work of estate management. Annotations on an anonymous map of lands belonging to Lord Spencer at Radbourne (Warwickshire) in 1634 show clearly how estate maps might be used over a long period (figure 2.15). In 1736 it was revised by Henry Beighton, who added a new scale and some new field names and areas. One or more other hands have pencilled subdivisions on some of the closes, adding numbers or "Not my Lords" or other notes, including some on land use, "Plow," "Probably plow," "Mow"; some more field names have been altered, and the line of a canal entered across the property. Finally, two hands of the late eighteenth century have added more notes on the closes and, at the bottom, a table listing the closes, their areas, and their tenants. All these annotations show that the map was in administrative use for at least 150 years. Estate maps were indeed made to last. Most were drawn on parchment, and on the written survey accompanying John Doharty's map of Dr. Thomas's estate at Atherstone-on-Stour in 1738 the name of the farmer is on a detachable slip that could be replaced without disfiguring the document; map and survey alike were intended to outlive the term of a single lease (figure 2.16).[44]

But to the landowner the estate map offered more than simple practical utility. Consider three late-seventeenth-century maps of the whole or

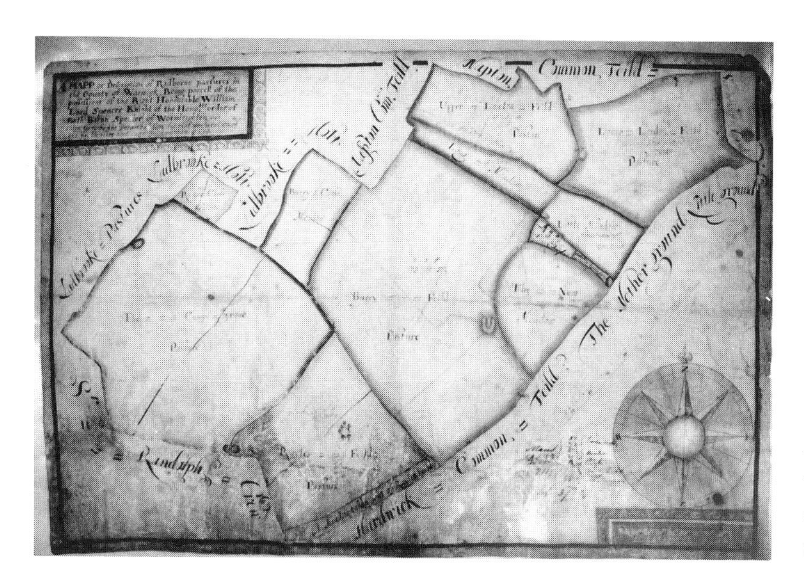

Fig. 2.15 Anonymous, lands belonging to Lord Spencer at Radbourne (Warwickshire), 1634 (Warwickshire County Record Office, CR 732)

An Exact Terrier to an Estate in the Parish of Atherston and County of Warwick Belonging to the Rev^d. D^r. Thomas. Done by Jo^n Doharty ju^r 1738

Tennant's Name	No. in Map	Names of the Grounds	Meadow A	Meadow R	Meadow P	Pasture A	Pasture R	Pasture P	Tillage A	Tillage R	Tillage P
	1	Mady-Fant Meadows	4	0	04						
	2		6	0	22						
	3	Fuzes				8	2	09			
	4	The Flat Grounds				8	2	20			
	5					11	1	25			
	6					11	3	14			
	7	Kings Close				11	2	29			
	8					9	3	07			
	9	Lor Hills							10	1	05
	10								9	0	24
	11	Robins Hill				10	3	07			
	12	The Sling				2	0	38			
	13	The Orchard				1	2	26			
Tho' Salmon	14	The Homestal				3	3	39			
	15	The Home Close				2	3	03			
	16	The Road within the Gates				1	3	17			
	17	Upper Had Leasow				6	1	00			
	18	Fithers				11	1	27			
	19	Dry Meadow				3	2	22			
	20	Middle Had Leasow				8	0	30			
	21	Rams Close				1	3	38			
	22	Preston Field Close				4	0	20			
	23	Lower Had Leasow				11	3	16			
	24	The Cow Ground				12	0	31			
	25	Atherston Meadow	2	0	16						
		In all A R P 173 2 13	12	1	02	111	3	22	19	1	29

Fig. 2.16 John Doharty, detail from the written survey accompanying the map of Dr. Thomas's estate at Atherstone-on-Stour (Warwickshire), 1738 (Warwickshire County Record Office, CR 936, box 3)

parts of Wedgnock Park (Warwickshire), which belonged to Lord Brooke. One, by James Fish in 1696, is a simple outline of three plots of land with their names and areas, uncolored and drawn on paper (figure 2.17). It is wholly functional and may indeed have been drawn for some immediate ad hoc purpose, not for general reference. Another, again by James Fish (figure 2.17) in 1696, is a more elaborate map of a mostly wooded area, a good deal more detailed, drawn on parchment and colored with gray wash—but still a functional and unpretentious map; there is no decoration, and the title and key are enclosed simply in borders of ruled lines (figure 2.18). The third, by John Wilson in 1682, is an altogether more splendid production figure 2.19). Drawn on two sheets of parchment glued together, it uses color for many features. In one corner is an elaborate cartouche for the title: a marble tablet with a nymph on either side and Lord Brooke's coat of arms with coronet and with two swans as supporters. Cartographically, none of these three maps is particularly complicated: mostly they show simple outlines of fields and

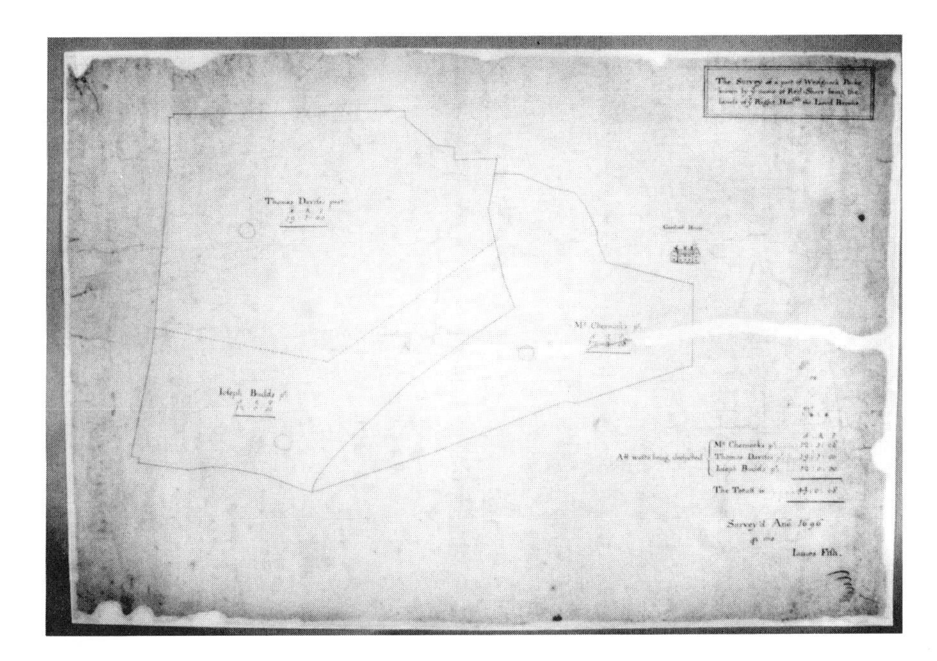

Fig. 2.17 James Fish, Wedgnock Park (Warwickshire), 1696 (Warwickshire County Record Office, CR 1886/M568)

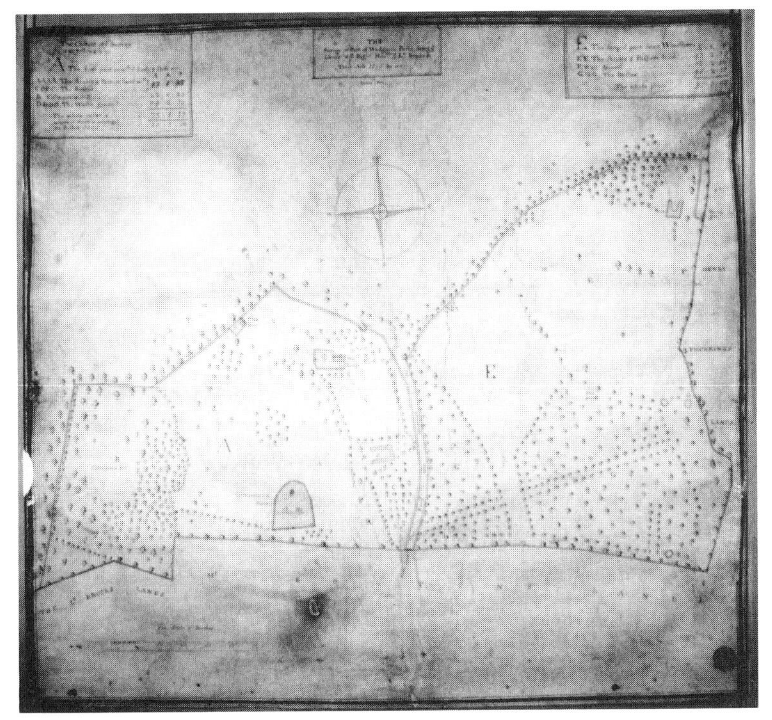

Fig. 2.18 James Fish, Wedgnock Park, 1696 (Warwickshire County Record Office, CR 1886/M2)

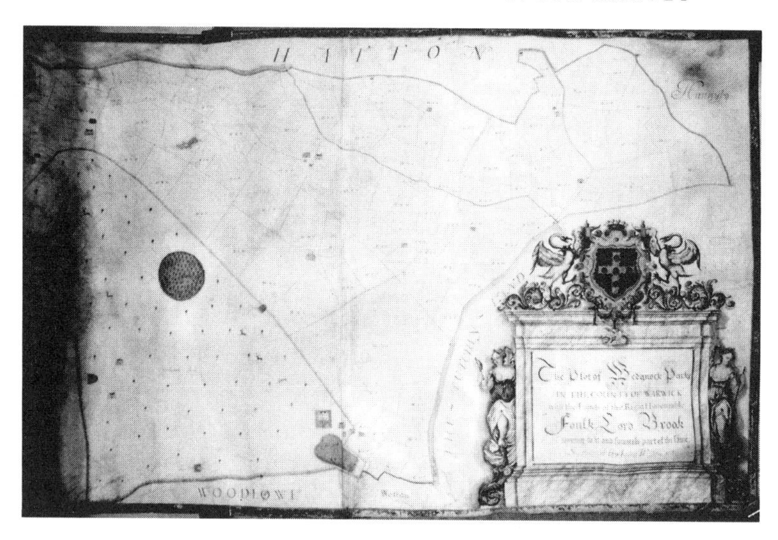

Fig. 2.19 John Wilson, Wedgnock Park (Warwickshire), 1682 (Warwickshire County Record Office, CR 1886/M3)

woods. The difference among them is in the presentation: the greater or lesser formality and ornamentation of the surrounds. There is no difficulty in finding other estate maps of every level of elaboration.

Clearly John Wilson's map of 1682 was meant for display—like many other estate maps. In the novel *Il gattopardo* (chap. 1), Giuseppe di Lampedusa's mid-nineteenth-century Sicilian Principe di Salina has a room decorated with mural pictures of the family estates, a graphic symbol and reminder of his power and wealth. In their more modest way many English estate maps served exactly this purpose. Nor need this be seen as a less practical, less functional purpose than the use of the maps in the detailed work of running the estate. They were meant to be looked at and to impress, to excite the beholder's admiration and to satisfy the owner's pride of possession without even the trouble of travelling to the spot. They were a statement of ownership, and of the status that came with ownership of land—and this was as true of Robert Smith with his 20 acres at Coundon as of Lord Brooke or the ruler of the kingdom.

This then was the value of English estate maps to those landowners who commissioned and paid for them, those who used and enjoyed them. They are of no less value to historians, to those of us who preserve them and who still use and enjoy them today.

The Historians and the Fields

Topographical history, the history of individual places, has been one of the great strengths of English historical scholarship from the time of William Lambarde, whose *Perambulation of Kent* was published in 1576, down to our own day, which has just seen the appearance of the 200th volume of the monumental *Victoria History of the Counties of England.* Central to this strength is the history of landed properties: how estates were built up and divided, how they passed from owner to owner across the generations, their size, their composition, their value. Legal requirements, family piety, antiquarian curiosity, and historical inquiry have over the centuries all contributed to the amassing of a vast amount of information about estates, incorporated in county histories and many other published works. However, it is only in the present century that this great tradition of research, with all its emphasis on the exact definition of rights, tenures, and amounts of land, has turned to precisely where properties lay, and thus to the sort of information that estate maps are peculiarly suited to provide. This is strange, for some of the tradition's early exponents were well aware of the value of maps and were even interested in the techniques of drawing them; a notable example is William Dugdale, whose maps of the hundreds were published in his *Antiquities of Warwickshire* in 1656.[45] With few exceptions it is only recently that topographical historians have begun to use estate maps systematically for the details of local landownership and local history.

This means that it was historians working on more general themes who led the way in using estate maps as a source of evidence, and the first to do so were the historians of field systems. The pioneer was Frederic Seebohm. In his book *The English Village Community*, published in 1883, he described the three-field system of cultivation that had operated before enclosure: the individual held his arable in strips, scattered among those of his neighbors, in open fields that were cropped following an agreed uniform rotation and were used as common pasture when they were not sown. He saw these arrangements as reflecting the economic and social pattern of the medieval and earlier periods. In the book's frontispiece is a map that is the direct ancestor of innumerable textbook and other illustrations of the three-field system: a map of Hitchin (Hertfordshire) in about 1816, before enclosure, showing the large fields that were the cropping units and the innumerable arable strips of the individual landholders. He took his map from one drawn to accompany a tithe apportionment, but if he had chosen another locality for his example, it could as easily have been from an estate map. Beside this map is another, also of Hitchin, showing the distribution of the strips and plots of land of a single landholder about 1750. The "Hand Map" (i.e., manuscript map) from which this was taken was probably an estate map.

Seebohm's interest in the way fields were organized before enclosure was followed up by many later historians. It was Seebohm who first discussed the historical implications of arable strip holdings in a three-field system. But the first historian to describe the variety of field systems in medieval England was H. L. Gray in his book *English Field Systems*, published in 1915. Maps accompanying tithe apportionments and estate maps were crucial to his work: he was in fact investigating the clear differences in field systems that early estate maps show. There was not only the three-field (or two-field) system that Seebohm described, the Midland system. There was the East Anglian system of holdings of strips and plots, concentrated in particular areas of open fields that were divided into so-called foldcourses for grazing sheep. There were the ancient enclosed fields of Kentish agriculture. There were the various combinations of closes with common fields held in strips, which were found in the other counties of southeast England. There was the system of heavily cropped infields and lightly cultivated rotating outfields, found in the areas of marginal cultivation beside upland moors. The outward differences among all these ways of organizing arable holdings leap to the eye on estate maps, and Gray made full use of them in his work. Presumably for clarity, he reproduced none in full, but used instead drawings taken from them; among them was Ralph Agas's 1575 map of West Lexham (Norfolk), already mentioned as the earliest English estate map that has so far come to light.[46]

Gray, like Seebohm, extended his investigations to the period after the Middle Ages, and many of the estate maps that Gray called in evidence were very much later than 1575. All the same, they provide invaluable information about what happened in the Middle Ages. It is not that nothing changed, that we can assume that in any one place the layout of fields was the same in the twelfth century as in the seventeenth. We cannot. The point is rather that the map is so clear and precise that it leaves open no question of the field layout at the time it was made; it thus acts as a key to a whole mass of contemporary and earlier written surveys and other documents from the same place and similar places. Without the map these records would be hard to explain and open to varying interpretations. The map explains them, and with their help we can carry back the picture the map gives into much earlier periods when estate maps were unknown. The earlier written records may show us that we have to modify the picture given by the later map; but at least we have a starting point firmly based in the local terrain for our understanding of the medieval fields of the place in question—and, by cautious analogy, of other places in the same area for which early estate maps are lacking. The map primes the pump, enabling us to use other evidence that would otherwise be obscure.

This use of estate maps as a key to written surveys was pioneered by Gray but has since been used by many historians in work on particular places and on English field systems in general. A notable example is the seminal work of C. S. and C. S. Orwin. The first edition of their book, *The Open Fields*, was published in 1938; in it they discuss the actual methods of agriculture—especially the techniques of plowing—that everywhere lay behind the strip cultivation of open fields. They then look at what happened in practice in one particular place, Laxton (Nottinghamshire), and use as a constant point of reference a magnificent estate map with accompanying written survey, revised by Mark Pierse in 1635 from a map and survey produced by Francis and William Mason ten years earlier. The map is an unusually fine one, very detailed and embellished with scenes of work in the fields, hawking, and hunting; the Orwins reproduced it (redrawn from the original) in full. With its help they explored other records, earlier and later, of the fields of Laxton, a place particularly suited for this study, since a form of openfield farming has, uniquely, continued there down to the present day. (Plate 3, from an early-seventeenth-century estate map, illustrates the making of hay.)

In all this the estate map is doing for the historian exactly what Ralph Agas in 1596 said it would do for the landowner: it records "bounders and abuttals" better than any written document can do, and it enables us to trace lands through changes of use or changes of name that have made the older written records difficult to understand. Moreover, Agas points out, once the information gathered by the surveyor has been presented in the form of a map, the map can be used to draw up a written survey: "upon the perfecting of any such surveigh, you may make a faire parchment booke with a large margent, you may enter, & ingrosse the same from the saide Plat."[47] This is perfectly true. But what Agas does not point out, though it is implicit in what he says, is that the opposite is not the case—and it is this that makes the map so valuable to the historian. In practice it is not generally possible to construct an accurate map from the information given in even the most detailed written survey. At first sight one might suppose it to be a fairly simple task. A survey like that of 1575 from Neithrop (Oxfordshire), which describes some two thousand strips in the fields, arranges them in the furlongs (or flatts, shots, or wongs: groups of parallel strips) in which they lay and, for every furlong, names the furlong, boundary, or other feature that adjoined it on each side.[48] A jigsaw puzzle to be sure, but one capable of patient solution. As F. W. Maitland put it in 1898, in writing of a mid-fourteenth-century survey that listed about a thousand pieces of land in Cambridge West Fields, "In each field it describes the various furlongs or shots in such a manner that an ingenious man, who had time to spare and a taste for the Chinese puzzle, might depict them on a map."[49] The passage conjures up a splendid picture of the Victorian gentleman, idly beguiling the time by playing with a medieval survey in this way. In fact he would have been lucky if he had succeeded, as anyone who has ever tried it will know. The survey gives bounds and areas, but it does not give angles or shapes, and few furlongs were rectangular. Fitting them with confidence into even the known facts of the local landscape can be all but impossible. Professor D. G. Kendall, who developed a computer program to achieve this construction of a field map from a written survey, suggests that if, as may happen, the survey gives on average only three of the four abuttals for each furlong, then the "margin of the possible" has been reached.[50] Where agrarian historians have a map, none of these problems arise: they start with the layout of the fields at one particular time laid out firmly before them and can interpret the written surveys and other records in its light.

The detailed layout of the fields is effectively information that only a map can give, and how they were worked is often clearer from a map than from any written source. There are many details of tenure and land use that surveys and other written records will provide but that are much more easily seen on an estate map. Varying measures are an example. A map by George Withers in 1623 of a property in Weston Patrick (Hampshire) gives the area of each parcel of the estate (figure 2.20). For agricultural land and woodland (marked with trees) only one amount is entered, but for copses (marked with alternate trees and shrubs) there are two, "lande measure" and "wood measure"; on one plot the former is just over 50.9 acres, the latter just over 42.5. The woods, as so often, were measured with a longer perch than the other lands—18 feet instead of the standard 16.5 feet. The copses, semi-woodland, were measured with both. The map displays graphically, in a way the written record cannot, the contrast between the two measures.

The same map by George Withers outlines the various plots of land in colors that are explained in a key: red (which over the centuries has changed to purple) is freehold land, green leasehold, and yellow copyhold. This again is information that a written survey could provide just as fully and accurately. But on the map we at once see the relative sizes and positions of the areas in the three different sorts of tenure, and therewith any implications for the layout of the estate, for its history, and for the history of the locality. Relationships leap to the eye, whereas painfully plotting on a map the information as given in a survey might be thought unlikely to be worth the trouble. This is often the case when a map distinguishes different parts of an estate. A map of Lord Brooke's manor of Cestersover, near Monks Kirby (Warwickshire), by James Fish in 1691 outlines the fields in color to show which lands belonged to each of four tenants; here again, the significance of the division of the four farms and their sizes and locations will at once be apparent to the historian working on the area.[51]

Sometimes, though, the coloring on an estate map raises more questions than it answers. William Folkingham, in 1610, described what colors to use for different sorts of land on a map and how to

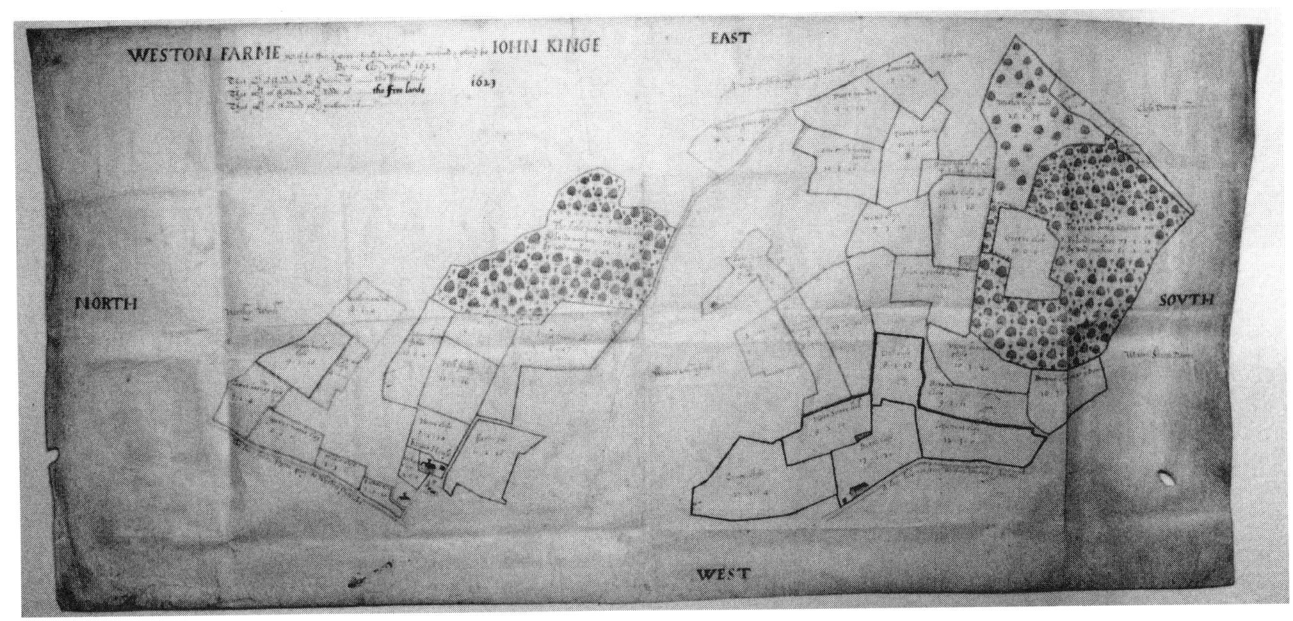

Fig. 2.20 George Withers, a property in Weston Patrick (Hampshire), 1623 (Hampshire Record Office, 11M 49/509)

make them. Thus "Arable for Corne may be dashed with a pale Straw-colour compounded of Yellow Oker, and White leade, or of Pincke and Verdigreece"; meadows were to be light green, pasture a deeper green, heaths and fens "deader Greene," trees "a sadder Greene," and so on.[52] It would be interesting to discover how closely contemporary maps conformed to this. There seem to have been few generally accepted conventions, and we may be at a loss to understand the colors on a map. Another Warwickshire map of 1691, of a farm in Butlers Marston by an anonymous surveyor, uses green, brown, yellow, and purple wash to color the fields; there is no key, and we can only guess what the colors show, perhaps differing land use.[53] Here the written record may explain the map, but it is far more often the other way around. To the historian of English fields and their exploitation, the estate map is a crucial source of evidence.

The Historians and the Landscape

The estate map's importance has been underlined by a more recent approach to agrarian and rural history, that of historians who use the landscape of today as evidence of the past and who at the same time uncover the processes by which that landscape has come into being. They follow the same technique as the historians of the fields who use early estate maps to throw light on other documents—but in their case these documents include the modern landscape itself. An early investigation of this sort concerned the ridge-and-furrow grasslands that are still found in many parts of England: former arable, now grassed over but retaining the shape of the broad ridges created for surface drainage when the land was last under the plow. These can be used to identify the detailed layout of the medieval fields, but with certainty only where other evidence confirms that this really was the ancient pattern—and the best evidence of all is an early map that shows the strips. Land normally under grass may have been plowed up for a short time in the Napoleonic Wars or at some other late date, so that the pattern of the ridges we see may be no older than the early nineteenth century.[54]

A pioneer in this work was M. W. Beresford. Especially important, not least for their use of estate maps, are his *History on the Ground* (1957) and, in collaboration with J. K. S. St. Joseph, *Medieval England: An Aerial Survey* (1958). In the first of these works Beresford reproduced the whole or parts of eight sixteenth-century estate maps, not redrawn as in the books of Gray and the Orwins, but direct from photographs. This was not, of course, an innovation. Full reproductions of estate maps had been made long before, to illustrate works on particular places[55] or in collections for their antiquarian or wider interest; J. L. G. Mowat's edition of *Sixteen Old Maps of Properties in Oxfordshire in the Possession of the Colleges of Oxford, Illustrating the Open Field System* was printed in 1888, in the wake of Seebohm's work. But what Beresford did for the first time was to combine the evidence of these maps with the evidence of other documents and of the landscape to show how places had changed over the centuries and how relics of their history can be found in the landscapes of today. The history of boundaries, of fields and villages, and of deserted sites, the creation of market towns and of gentlemen's parks, were all viewed with the help of these early estate maps. In all this Beresford was using the maps as a way into past landscapes, a way of seeing what places really looked like when the maps were made.

This, of course, they do supremely well and they can thus be used, in a way we can use no written source, to trace change in landscape. We need look no further for examples than maps showing field systems before enclosure; but there are many others. Windmills are a feature of the rural landscape and economy that may be easily known only through their appearance on estate maps: they were apt to be short-lived, and the sites they occupied might be chosen from a great variety of suitable places on any one property. Thus one at Wykham, outside Banbury (Oxfordshire), appears as a landmark on one of the road strip maps in John Ogilby's *Britannia* in 1675 and again on an estate map of 1688, which is our only evidence for its precise location; as it does not appear on an estate map of 1746, we can be fairly sure that it had vanished by then. Without these maps all knowledge of the mill's existence would have been lost, apart from records of Windmill Quarter and Windmill Grounds as local field names.[56] Occasionally a map shows a settlement that has disappeared, even marking the site of a village or hamlet already abandoned when the map was made. A classic instance discovered by Beresford is on Thomas Clerke's map of Whatborough (Leicestershire), drawn for All Souls College in 1586, which names an area in the center of the map "The place where ye town of whateborough stoode"; the last houses of the village had been abandoned a hundred years earlier.[57] More often we are shown houses that were standing when the surveyor made the map but that have since been deserted. A map of Coughton (Warwickshire) in 1695 shows four houses, with their plots of land behind, in an area that was later taken into the grounds around the neighboring Coughton Court; on the ground, only the raised areas where the houses stood are now visible. Another now vanished feature on the same map is the moat around Coughton Court itself.[58]

Sometimes maps provide evidence of the actual process of change in the landscape. Beresford reproduces and discusses two maps of Holdenby

(Northamptonshire), which show how, between 1580 and 1587, the grounds around the mansion were developed and some 600 acres of arable, meadows and woods were converted into a park.[59] Both are in the volume of surveys and maps of Sir Christopher Hatton's estates drawn up by Ralph Treswell, and were presumably drawn first to plan and then to record the changes—even as a deliberate "before and after" comparison. Two other maps in the same volume record change even more precisely (figures 2.21 and 2.22). They show lands at Kirby, near Corby (Northamptonshire), in 1585 and 1586; between the two there had been enclosure and an exchange of open-field arable, recorded on the maps by the disappearance of many strip holdings and by details of what had happened entered in a panel on the second map. This apparently intentional record of change is most unusual—it was only the richest landowner who could afford to be so prodigal with his surveyor's time and skills. But successive maps from a single estate quite often demonstrate the sequence, even the process, of change, and even a single map need not be a static document, a record of a single point in time. We have already seen how the map of Radbourne (Warwickshire) in 1634 has annotations in a series of different hands, showing that it was being used in managing the property down to the late eighteenth century. These notes and alterations themselves record detailed change: the subdivisions on some of the fields, notes on the way the land was to be used, and some new field names, such as Barn Ground on Lower London Field—we remember Agas's comment that a map, unlike the written survey, avoids

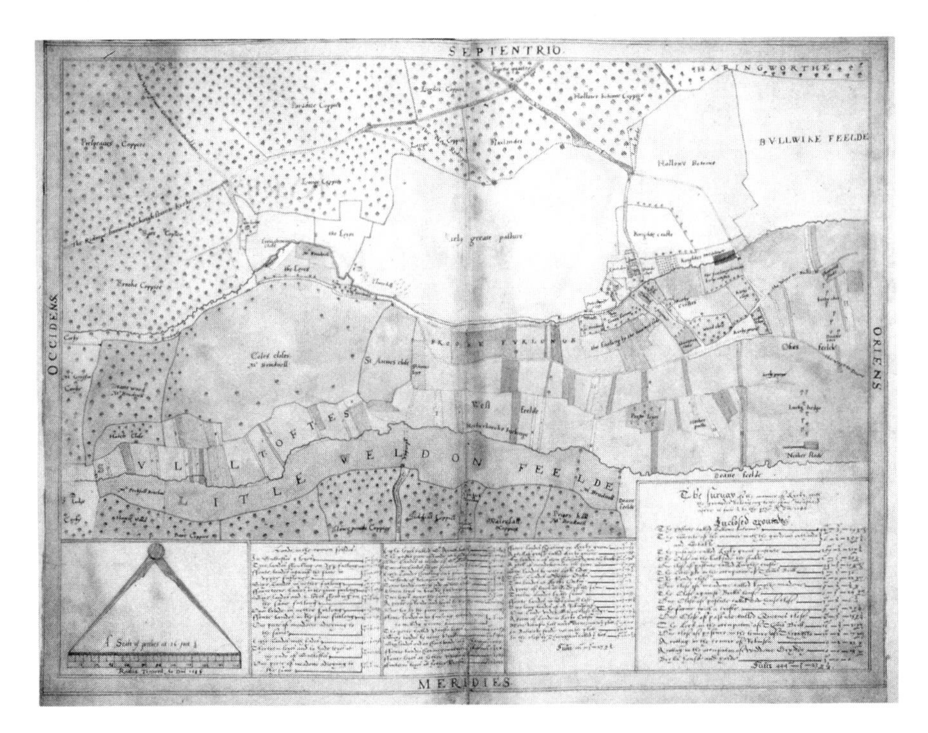

Fig. 2.21 Ralph Treswell, lands at Kirby (Northamptonshire), 1585 (Northamptonshire Record Office, Finch Hatton MS 272, fol. 5)

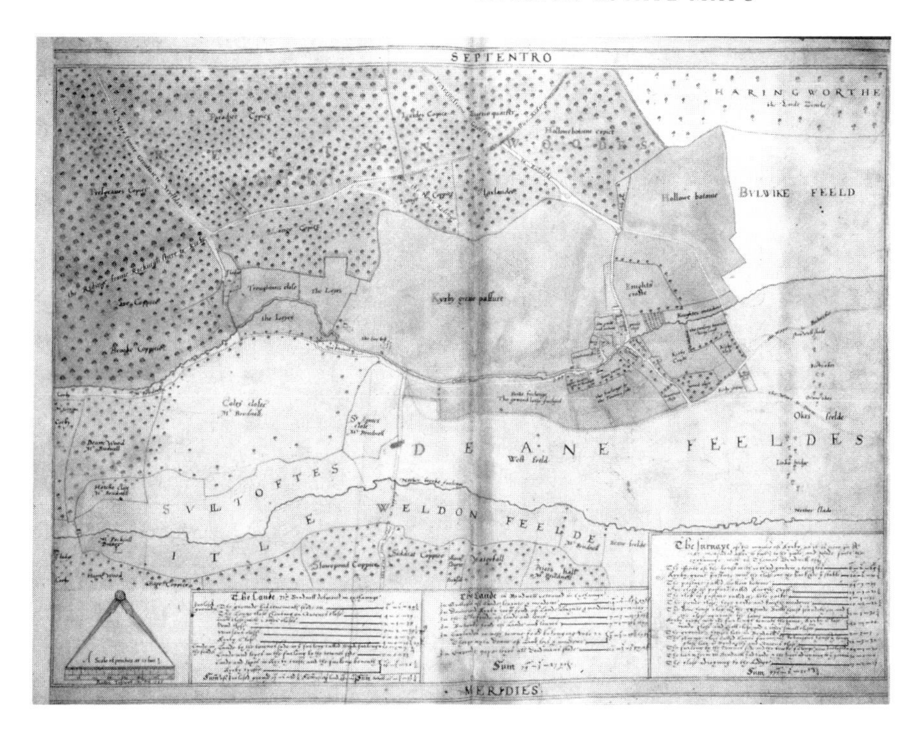

Fig. 2.22 Ralph Treswell, lands at Kirby, 1586 (Northamptonshire Record Office, Finch Hatton MS 272, fol. 6)

problems caused by changes of name. Correctly interpreted, this one map can tell us much of the history of the property over a century and a half.

Besides providing a record of change the map places the past landscape before us with an immediacy and comprehensiveness equalled by no other source. No other mapmaker left quite so little to the imagination as Thomas Hornor, who in the early nineteenth century produced estate maps on a system that he called "panoramic chorometry." On the map was superimposed a near-vertical view of every feature, yet the map as a whole is not the bird's-eye view that it seems at first sight, for the basis was a true ground plan with consistent scale, and each feature is a separate perspective picture, drawn to a separate horizon. The result is a kind of isometric aerial view, detailed and even beautiful; but it was clearly slow and costly to produce. The same technique was used by some artists in the sixteenth and seventeenth centuries for engraved views of cities, but Hornor's attempt to apply it—in full color—to estate maps seems to have been unique.[60] All the same, some of the finer eighteenth-century surveyors approached fairly close to Hornor's method in the amount of detail on their maps and in showing it in near-vertical actual pictures instead of conventional symbols.[61]

Given the ability to read a map and a modicum of imagination in interpreting it, we can gain almost as vivid a picture of past landscape from many estate maps. For examples let us look at two of the most remarkable of the English estate maps of the sixteenth and seventeenth centuries.

The map of Lord Cheney's manor of Toddington (Bedfordshire) was drawn by Ralph Agas in 1581.[62] It is on twenty pieces of parchment, which have now been separated but which together form a map that measures 11.3 by 8.5 feet. Starting at the village, we see that the church and all the houses are drawn not conventionally but from life, in some detail, as though in oblique bird's-eye view. We should not rely on the map's evidence for all minutiae, such as the position of every door and window; but we need not doubt the accuracy of the overall picture, for not only does Agas's work show every sign of meticulous care, but it was done for those who would have known intimately every nook and corner of the place. We see the market cross and the maypole and get an impression of the sizes, shapes, and general appearance of the houses and of the distribution of trees (figure 2.23). From what other source could we get so detailed a picture of an Elizabethan village street? Moving beyond the village area, the map gives us a superb view of the agrarian landscape. Arable strips and closes, meadows, woods with deer in them, trees and hedges along the boundaries are all shown. There is of course a conventional element—Agas cannot be expected to show us the exact position of every tree, and he is only telling us that this is where trees were to be found. But his representation is wholly pictorial, so that one can easily imagine oneself walking along the paths, seeing what one would see on the ground; in some ways, indeed, one sees rather more on the map, for it identifies the area, the tenant, and the form of tenure of every plot of land. Moving to another part of the map, we see Lord Cheney's mansion in bird's-eye view, clear enough to show its detailed architecture; it was a three-storied house, with gatehouse and turrets, built around a central courtyard with another large courtyard behind (figure 2.24). A broad avenue of trees leads up to the yard in front of the house, and beside the house are formal gardens and an orchard.

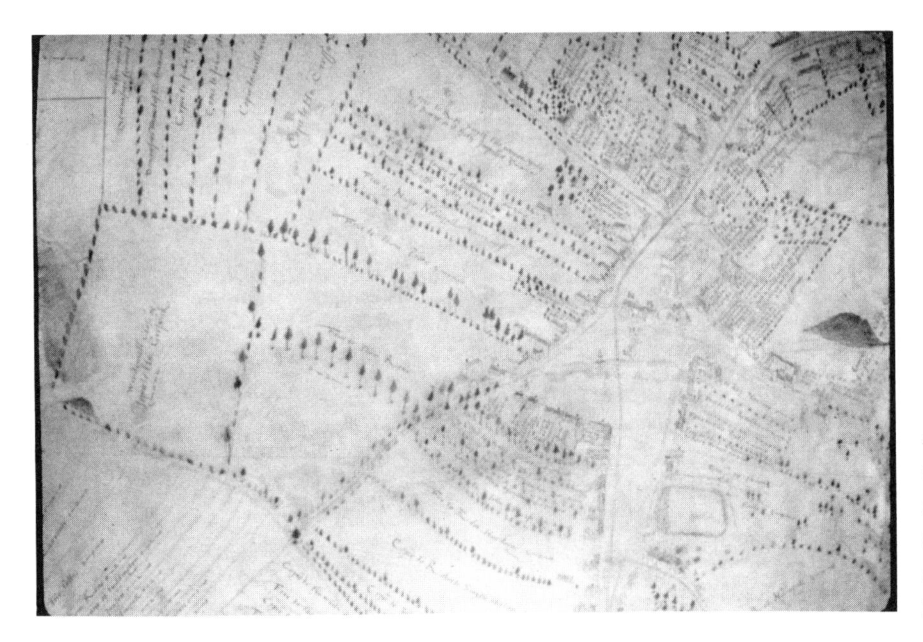

Fig. 2.23 Ralph Agas, detail from map of the manor of Toddington (Bedfordshire), 1581 (British Library, Add. MS 38065)

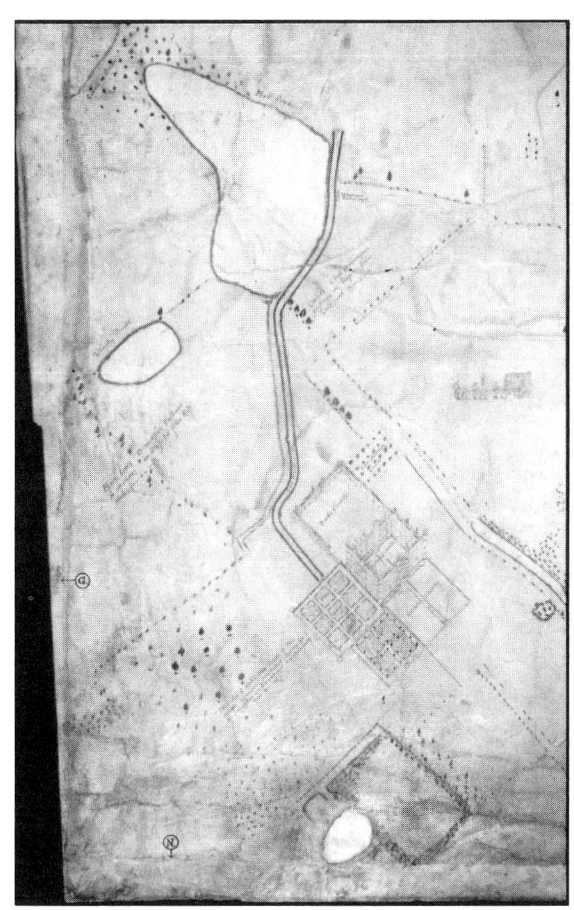

Fig. 2.24 Ralph Agas, detail from map of the manor of Toddington

Beyond are areas of water, the Black Pond and the Mill Pond, and a loaded two-wheeled cart with a team of horses crossing a field—to the cartographer a mere embellishment, but to the historian a further source of evidence. The whole map provides a mass of information on—quite apart from anything else—what Toddington looked like 400 years ago.

We see this again in the map that James Fish produced in 1692 of the earl of Rochester's estate at Kenilworth (Warwickshire).[63] Drawn on six pieces of parchment glued together, it is much smaller than the Toddington map, but measuring some 6 feet square, it is still an ambitious and impressive work. In the lower left corner is Kenilworth itself, a small town rather larger than Toddington and spread out spaciously over quite a wide area. Beside it is a panel giving a key to the names of its streets: New Way, Pepper Alley, Rosemary Hill, and so on. The scale is smaller than on the Toddington map, but we still see a great deal of detail: individual houses with their red roofs, what are probably inns with signs outside, the church with its steeple and a weathercock on top. The smaller houses are drawn conventionally, but the larger ones may well be from life. At one end of the town is the castle. On the map itself it appears within its outline site drawn in a weird semiperspective; but beside the map are two more conventional pen-and-wash views of the castle, one from each side, with a horse and cattle in the foreground. This marks a new fashion, common in the eighteenth century, of embellishing estate maps—or at least those intended for display—with separate views of parts of the property. Those of Kenilworth Castle are especially interesting because some forty years earlier, after the Civil War, it had been slighted—made unusable for military purposes—and the great pool or mere beside it drained. The map marks the outline of where the pool had been, and the views of the castle show how little visible damage the slighting had produced—the castle was in fact still used as a residence (figure 2.25). They also show its vast size; during the slighting and draining Colonel Hawkesworth, in charge of operations, had made a home for himself in the gatehouse alone.[64]

Beyond the castle we move into the countryside, and the map again gives us an extraordinarily detailed picture of what was to be found. Roads

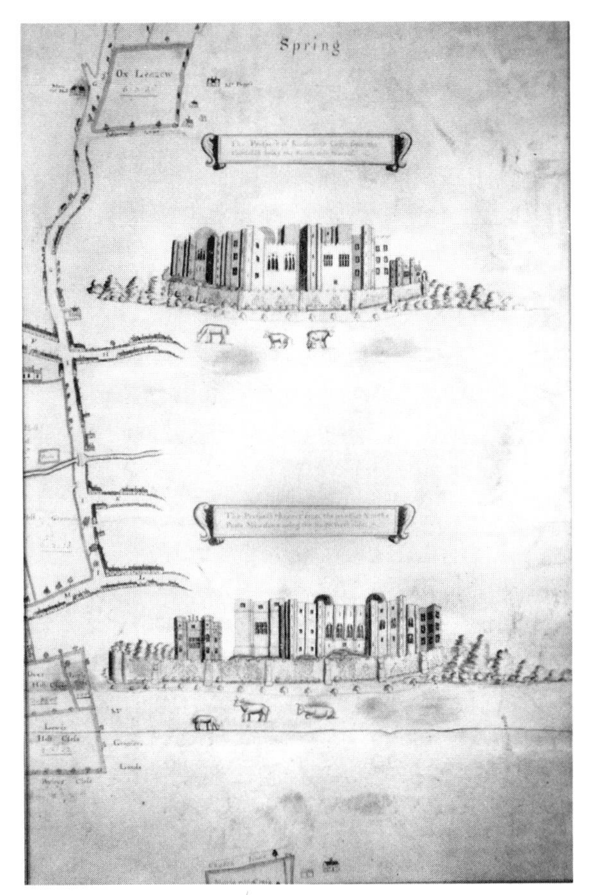

Fig. 2.25 James Fish, detail from map of Kenilworth (Warwickshire), 1692 (Warwickshire County Record Office, CR 143A)

with trees and hedges on either side, streams crossed by bridges or fords, ponds, gates into the fields—all of them individually owned closes, for unlike Toddington a century earlier there were no common fields with strip holdings in late-seventeenth-century Kenilworth. One has to look closely at the smallest details of the map to realize how much it tells, how from every point one can envisage what one would actually have seen on the ground. In the farther parts of the estate were woodlands, marked with trees that are no more than conventionally placed but that reinforce the overall pictorial quality of the map. Here and there in the landscape are isolated farms, with their gardens and orchards beside them, and around them are the lands belonging to them—as we could not tell on the ground but can see easily enough on the map, for every tenant farmer's lands are outlined in a different color. Nor could we see on the ground what is very clear on the map, the area and name of every field: Coneyberry Hill, Teutawing Close, Boggy Meadow. In fact the survey book accompanying the map tells us practically nothing the map does not except the rent paid to the earl by each farmer.

One could write a whole book about what is shown on each of these two maps, of Toddington and Kenilworth. Both are highly pictorial, and one needs little imagination to tell what one would see as one walked along every road and path: every cottage and outbuilding, every stream, gate, row of trees, and so on. Because in imagination one can move around the area in this way, the map sets us free in the past landscape in a way that a painting, which offers vivid realism but is tied to a single viewpoint, does not. The maps of Toddington and Kenilworth are of course exceptional works—but only in their size and comprehensiveness. Their accuracy and their detail can be paralleled on many smaller, less ambitious productions. Properly viewed, almost any estate map can give one insight into past landscape.

Moreover, once one becomes accustomed to the genre, once one gets one's eye in, so to speak, almost every estate map has some peculiarity, some unusual feature of particular interest. An example is the map of Brixworth (Northamptonshire), drawn by the vicar, Richard Richardson, in 1688. Unlike most estate maps it includes notes of the types of soil in different parts of the area: "Red Land" and "Clay Land." An even stranger feature is that it shows the standing places of ricks, marked with green-outlined rectangles. Again, Richardson was clearly interested in natural history, and the map includes notes on the wildflowers: "Violae primulae," "Serpyllum his locis nascitur" (i.e., thyme grows in these places), and so on.[65] Of course this is not a typical estate map—but very few estate maps are. There is something unexpected and interesting to be found on almost every one.

The Historians and the Landowners

When we use any historical document, we must always ask ourselves why it was written, for this was almost certainly not to give us information about the age that produced it but for some immediate, practical purpose that affected its form and content. If we treat it simply as an objective data source, we are likely to be led astray; we must make use of what it tells us in the light of what we believe it was trying to do. This is as true of estate maps as of any other sort of record. They are not general topographical maps like the official large-scale maps of today. They may incidentally show us how fields were cultivated before enclosure, what past landscapes looked like, and a host of other things, but this was not why they were drawn. Even when we use them for so seemingly neutral a purpose as a reconstruction of the landscape, we must be on our guard. The map was drawn for the landowner who, for one thing, is bound to have expected from his surveyor a reasonably flattering picture of his property. We may expect to find on the map the castle, the church, the neat row of houses on the village street, the hedges and fences carefully kept. What we cannot expect to find but must supply from other sources of evidence—even our own imagination—are the ruts and boggy patches in the roads, the sorry disrepair of the bridges and gates, the derelict outbuildings, the dung heaps.

Moreover, the map was usually drawn to show nothing except the property of the landowner. Sometimes this limitation is obvious. We see it at once in a map drawn by Thomas Hill in 1681 for the dean and chapter of Canterbury (figure 2.26). It shows their property at Vauxhall (Surrey), beside the River Thames: a series of closes in the hands of various tenants, together with houses lying along streets that are shown enlarged in a special inset. Beyond the area where their land lay, the map continues but shows only outstanding features, apparently included just to set the property in its local context: Spring Garden, Lambeth Palace and the row of houses joining it to Vauxhall, and on the other side of the river the houses of Westminster and the abbey behind. It is very clear that these features have all been drawn impressionistically: they form no part of the surveyed map. But much less clear is the status of closes belonging to other landlords that lay interspersed among the Canterbury lands; their owners are named, but their areas are not given. If the surveyor established the correct relative positions of the dean and chapter's pieces of land, he must have made some measurements of these intervening closes as well; but we cannot be sure that this was done, still less that the bounds between these other lands have been even sketched in from the ground—it could well be that they were put in simply to give the map an appearance of completeness and bore no relation to reality at all. This was the normal practice in entering neighboring properties on a map of an estate that formed a single block of land: roads and houses might be sketched in approximately the right positions, but field boundaries would probably be merely drawn from imagination to produce a more

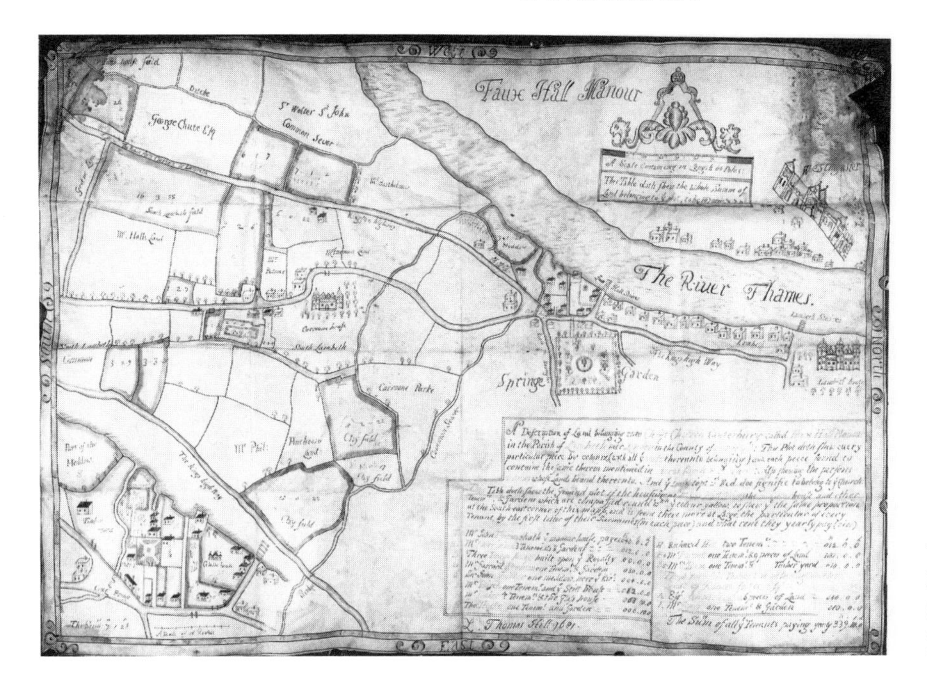

Fig. 2.26 Thomas Hill, property at Vauxhall (Surrey), 1681 (British Library, Add. MS 34790)

or less convincing pattern. This is almost certainly, for instance, the procedure followed on John Norden's maps of the Honour of Windsor in 1607 (figure 2.13).

Bearing in mind, then, why estate maps were made, it follows that they will tell the historian most reliably about the estates themselves: how they were run and how they were perceived by their owners. We have already seen that a map will often have quite a different archival history from the other records of an estate. Because they are picturesque or quaint, or even beautiful, estate maps are often divorced from the context of the other records, even from the written surveys that they were meant to accompany. They may have been hung up as decorative pictures; they may have been preserved when other records were thrown away, or sold when other records were kept. Thus almost all the estate maps in the British Library were acquired as single items, whether by gift or through the sale room. This means that we see them apart from the other records created in running the estate and cannot appreciate what part they were meant to play, how they were used in the work of management. The historian cannot, in other words, effectively draw on this important selection of estate maps as evidence for one aspect of the past on which they are peculiarly fitted to give information; and it is just that aspect on which we need information if we are to understand precisely why and how estate maps came to be made. Elsewhere, however, plenty of maps survive within their original archival context, often, indeed, still in the hands of the estates for which they were created. Sarah Bendall, in her work on the Cambridgeshire maps, is leading the way in investigating the role of estate maps in actual estate administration. This may well produce results of great interest.

On the other hand, it need not be unhistorical to set the map apart from the other records of an estate. It seems unlikely that when the earl of Rochester got his map and survey book of Kenilworth from James Fish in 1692 he handed them over to his steward to keep along with the accounts, the leases, the rent rolls, and all the other estate papers. What such a map as this has to tell us is not so much how the landowner ran his estate as how he perceived it. We should see it as part of the baroque pomp with which the great surrounded themselves at this time, at one with the Palladian mansions, the liveried servants, the funeral monuments. Nor need it be only the great whose estate maps should be looked at in this way. A lesser man might view his estate map rather as someone today might view his bank statement when it shows a healthy balance. It was not for administrative efficiency that another Warwickshire landowner, Robert Smith of Coundon, had his 20-acre property mapped in 1694 (figure 2.14). This too has something to tell the historian.

Notes

In these notes the following abbreviations are used: BL (British Library), PRO (Public Record Office), WCRO (Warwickshire County Record Office).

1. The first few pages, an account of the first appearance of English estate maps in the 1570s, are abridged from P. D. A. Harvey, "Estate Surveyors and the Spread of the Scale-Map in England, 1550–1580," *Landscape History* 15 (1993): 37–49. P. D. A. Harvey, *Maps in Tudor England* (Chicago, 1993), 79–93, sets estate maps in the context of the general development of mapping in the sixteenth century.

2. BL Add. MS 38065.

3. Somerset Record Office, DD/MI, c/186; Holkham Estate records, map 87a (photocopy, BL Maps 188 n.1 [10]); Guildhall Library, London, MS 25517/1. An earlier map that fits our definition of an estate map was drawn between 1444 and 1446 to show Edmund Rede's manor of Boarstall (Buckinghamshire), but this was an isolated production without successors (R. A. Skelton and P. D. A. Harvey, eds., *Local Maps and Plans from Medieval England* (Oxford, 1986), 211–19, plate 18). See color plate 58 in Harvey, *Maps in Tudor England,* for a map by Israel Amyce.

4. Northamptonshire Record Office, Finch Hatton MS 272. Israel Amyce's survey with maps, made in 1579, of the lands in Essex of the late Edmund Tirrell (BL Harley MS 6697) scarcely meets our definition of an estate map, for it was made so that the lands could be divided between Tirrell's heirs.

5. Skelton and Harvey, *Local Maps and Plans,* 71–81, plate 3.

6. W. Ravenhill, "The Plottes of Morden Mylles, Cuttell (Cotehele)," *Devon and Cornwall Notes and Queries* 35, pt. 5 (spring 1984): 165–74, 182–83.

7. PRO MPC 77, fols. 4–7.

8. PRO MR 2.

9. L. R. Shelby, *John Rogers: Tudor Military Engineer* (Oxford, 1967); P. D. A. Harvey, "The Portsmouth Map of 1545 and the Introduction of Scale-Maps into England," in *Hampshire Studies,* ed. J. Webb, N. Yates, and S. Peacock (Portsmouth, England, 1981), 32–49; M. Merriman, "Italian Military Engineeers in Britain in the 1540s," in *English Map-Making,* ed. Sarah Tyacke (London, 1983), 57–67.

10. PRO MPF 137; R. A. Skelton and J. Summerson, *A Description of Maps and Architectural Drawings in the Collection Made by William Cecil, First Baron Burghley, Now at Hatfield House* (London, 1971), nos. 73–76, 78, 80.

11. D. Hodson, *Maps of Portsmouth before 1801* (Portsmouth, England, 1978).

12. PRO MPF 240; see Harvey, *Maps in Tudor England,* color plate 22.

13. PRO MPF 257, MPF 144; G. Manley, "The Earliest Extant Map of the County of Durham," *Transcations of the Architectural and Archaeological Society of*

Durham and Northumberland 7 (1936): 278–87.

14. R. A. Skelton, *County Atlases of the British Isles, 1579–1703* (London 1970), 7–16.

15. Ralph Agas, *A Preparative to Platting of Landes and Tenements for Surveigh* (London, 1596), 16.

16. Skelton and Harvey, *Local Maps and Plans*, 12–18.

17. BL Lansdowne MS 165, fol. 91.

18. John Norden, *The Surveyors Dialogue* (London, 1607), 17.

19. Privately owned (copy, Dorset Record Office, photocopy 500).

20. PRO MPC 245, MPC 244(1), MPC 91.

21. PRO MPC 245. The document tells how those commissioned to correct an earlier map, found to be faulty, had discovered it was so badly wrong that they had to draw a new one; one of their number withdrew from the operation because they were excluding from the survey a particular piece of land not specifically named in their commission. This piece of land in fact appears on the map—unavoidably, given its position in relation to the pieces specified. It follows, though this is not made clear on the map itself, that only the specified pieces were actually measured and plotted to scale. This is confirmed by the speed of the operation and by the map's sketchy appearance overall.

22. Winchester College Muniments, 21443; 21445 (dorse) is a contemporary copy.

23. Winchester College Muniments, 21444a,b, 21447a–f.

24. WCRO CR 936, box 15/7.

25. I. M. Evans and H. Lawrence, *Christopher Saxton: Elizabethan Map-Maker* (Wakefield, England, 1979), 79–80, 122.

26. WRCO, card index of surveyors.

27. BL Maps MT 6 b.1 (30).

28. BL Add. MS 38065.

29. Borthwick Institute of Historical Research, York, PR.BP 15.

30. WCRO CR 143, CR 143A.

31. WCRO CR 936, box 3.

32. Hampshire Record Office, 23 M 54/1.

33. WCRO CR 285/3, 56. The survey is dated April 1598, forty-first year of Elizabeth I; but April 1598 was in the fortieth regnal year. The date of the court entered immediately above the survey is Thursday, 19 April in the forty-first year, certainly 1599, when 19 April fell on a Thursday.

34. WCRO CR 285/57.

35. The map and survey book from Fish's survey of Kenilworth in 1692 (above, n.29) are a good example. The map was deposited in the WCRO in 1950 by the earl of Clarendon, descendant of the earl of Rochester for whom the survey was made; the book came to the WCRO four years later from a different source.

36. WCRO L 6/1032.

37. The information for Warwickshire, collected primarily by D. J. Pannett, is available in a card index in the WCRO; the project is described, with an interesting and informative account of the history of estate maps in the county, in D. Pannett, "The Manuscript Maps of Warwickshire, 1597–1880," *Warwickshire History* 6, no. 3 (1985): 69–85. G. A. Rushton is collecting the information for Hampshire for publication in the Hampshire Record Series, and I am most grateful to her for making it available to me.

38. The one County Durham map is of Layton by Robert Farrow in 1608. It is in the Baker Baker papers in the University of Durham, Department of Palaeography and Diplomatic, 72/249, enclosure attached to fol. 12.

39. P. Eden. "Three Elizabethan Estate Surveyors: Peter Kempe, Thomas Clerke, and Thomas Langdon," in Tyacke, *English Map-Making*, 68–84; C. M. Woolgar, "Some Draft Estate Maps of the Early Seventeenth Century," *Cartographic Journal* 22 (1985): 136–43.

40. D. M. Barratt, ed., *Ecclesiastical Terriers of Warwickshire Parishes* (Dugdale Society, 1955–71), 1:ix–xii.

41. University of Durham, Department of Palaeo-

graphy and Diplomatic, DR terriers, Stanhope 1663; Hampshire Record Office, 21 M 65/E15/132. I am grateful to G. A. Rushton for information about the maps among the Hampshire glebe terriers.

42. E.g., G. Jacob, *The Complete Court-Keeper; or, Land-Steward's Assistant,* 8th ed. (London, 1819), 143–51. Cf. P. D. A. Harvey, *Manorial Records* London, 1984), 61.

43. Norden, *Surveyors Dialogue,* 83–144.

44. WCRO CR 1945.

45. P. D. A. Harvey and H. Thorpe, *The Printed Maps of Warwickshire, 1576–1900* (Warwick, 1959), 11–15.

46. H. L. Gray, *English Field Systems* (Cambridge, Mass., 1915), 317.

47. Agas, *Preparative to Platting,* 14–15.

48. Bodleian Library, Oxford, MS Top. Oxon. c 454.

49. F. W. Maitland, *Township and Borough* (Cambridge, 1898), 56.

50. D. G. Kendall, "The Recovery of Structure from Fragmentary Information," *Philosophical Transactions of the Royal Society of London,* ser. A, 279 (1975): 559–75.

51. WCRO CR 1886/M12.

52. William Folkingham, *Feudigraphia: The Synopsis or Epitome of Surveying Methodized* (London, 1610), 57.

53. WCRO CR 2103/1.

54. M. W. Beresford, "Ridge and Furrow and the Open Fields," *Economic History Review,* 2d ser., 1 (1948–49): 34–45; E. Kerridge, "Ridge and Furrow and Agrarian History," *Economic History Review,* 2d ser., 4 (1951–52): 14–36.

55. E.g., the map of Dewsbury (West Riding of Yorkshire) by Christopher Saxton, 1600, in S. J. Chadwick, *Dewsbury Moot Hall: Account Rolls and Court Roll* (Leeds, 1911), facing p. 1.

56. *The Victoria History of the Counties of England: Oxfordshire* (London, 1972), 10:71.

57. M. W. Beresford, *History on the Ground* (London, 1957), 95, 116–23, plate 11.

58. WCRO CR 1998/M7A, B. I am grateful to M. W. Farr for this information.

59. Beresford, *History on the Ground,* 211–14, plates 19, 20.

60. R. Hyde, "Thomas Hornor: Pictural Land Surveyor," *Imago Mundi* 29 (1977): 23–34.

61. E.g., the engraved map by John Rocque of the gardens of Richmond Old Park, 1734 (copy, BL Maps K.41.16f).

62. BL Add. MS 38065.

63. WCRO CR 143A; the accompanying survey book is CR 143.

64. P. K. B. Reynolds, *Kenilworth Castle, Warwickshire* (London, 1948), 3–4; *The Victoria History of the Counties of England: Warwickshire* (London, 1951) 6:137–38.

65. Northamptonshire Record Office, map 1555.

Chapter **THREE**

ESTATE MAPS *of an* ENGLISH COUNTY: CAMBRIDGESHIRE, *1600-1836*

SARAH BENDALL

Display of Estate Ownership

Visitors to stately homes in British Isles who keep their eyes open will frequently see paintings and engravings of the owner's estates; on occasion, there may be splendid maps of the property displayed. One can hardly fail to be impressed by one of the earliest, and very fine, examples of an estate map on show to the public: that of the property of the London lawyer and master of the rolls, Sir William Cordell, at Long Melford in Suffolk. The map was drawn by Israel Amyce (or Ames) in 1580: one of the foremost surveyors of his day, Amyce was living at Barking in Essex at the time that he was employed by Cordell; he surveyed and mapped estates for other private landowners and the Crown between 1576 and 1607 in the counties immediately north of London and in East Anglia.[1] Hanging in Long Melford Hall, the map is magnificent: it is drawn on nine sheets of vellum, is 100 inches high and 79.5 inches wide, and shows a wealth of detail, at about 15 inches to the mile, about the contemporary village, buildings, and land use. The table underneath the plan lists tenant holdings, fields, and acreages.[2] Having enjoyed the map and marvelled at its splendour, the visitor may proceed from the hall to the parish church to see Sir William's impressive Renaissance monument.

Even if maps are not on public display, there is a good chance that they exist among the estate papers of the aristocracy and landed gentry. Many instances can be found in Cambridgeshire. Cheveley Park has long had connections with the horse-racing community at nearby Newmarket. Henry Jermyn, Lord Dover, acquired the Cheveley estate in 1671 and had a new house built for him, in the latest fashion.[3] He was often visited by royalty when they stayed at Newmarket Palace for the races. Jan Siberechts, one of a number of Dutch émigrés who was commissioned to paint prospects of English country houses,[4] did so for Cheveley in 1681. The painting (figure 3.1) documents the newly built house, flower garden, park, stables—all a hive of activity. Thus the owner's pride in his new home and his ability to employ a prestigious artist to record it are demonstrated for all to see. In the early eighteenth century the estate was bought by

Fig. 3.1 Jan Siberechts, Cheveley Park (Cambridgeshire), 1681 (John Harris, *The Artist and the Country House: A History of Country House and Garden View Painting in Britain, 1540–1870* [London, 1979])

the sixth duke of Somerset, and as part of his alterations to the house and grounds, a plan was drawn in about 1735. This map, on paper, is a working document: it shows old and new foundations, walls, the depth of a pond, and "An Old Pond th*a*t wont hold Water Which is began upon to be fill*e*d vp with Brick Rubbish, &c."[5] In 1750 Somerset's daughter, Lady Frances Seymour, married John Manners, marquis of Granby, and the estate passed to him. His son Charles inherited the property in 1770 and succeeded as fourth duke of Rutland in 1779.[6] Known as an "amiable and extravagant peer, without any particular talent, except for conviviality,"[7] he, too, had his newly acquired estate mapped, in 1775 by Thomas Warren of Bury St. Edmunds in Suffolk (figure 3.2). Son of a schoolmaster and land surveyor, he and his brother Bendish inherited their father's surveying instruments. Thomas, too, had a school and in 1809 won an award from the Society for the Encouragement of the Arts for designing an improved school slate.[8] Warren's map is another high-quality piece of work. Drawn on

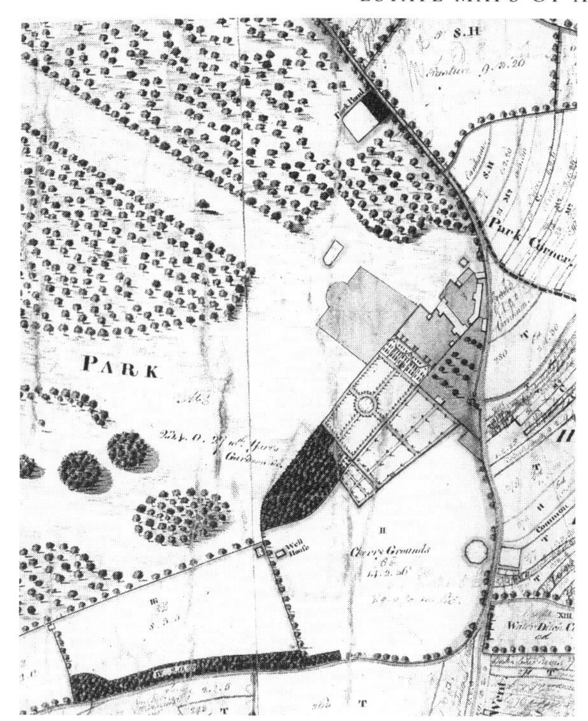

Fig. 3.2 Thomas Warren, part of estate map of Cheveley, 1775 (County Record Office, Cambridge, 101/P/2)

parchment at 4 chains to the inch and 68 inches high and 50 inches wide, it shows farms, almshouses, a well, a post mill, an inn, cottages and a warren house in perspective, Cheveley Park itself in exaggerated perspective, outbuildings and a brick kiln in plan form, and an ornament seat for watching horse racing, "The King's Chair," on the heath. The map has a decorative title cartouche, scale, and compass rose, and is accompanied by a terrier. Here then is another map that was drawn to impress and for display, and that occurs in a sequence of depictions of the estate.

How typical are these maps, how should they be interpreted, why were they drawn, how were they used, who made them, and who commissioned them? These are some of the questions that arise. One way of exploring them is to study a set of maps of a particular area, to compare and contrast them, and to examine them in relation to contemporary papers and documents.

Presented here is a summary of the findings from one such study, of rural estate maps of historic Cambridgeshire to 1836 (the date of the Tithe Commutation Act, which caused many areas to be mapped at similar scales to estate maps and thus influenced later mapping).[9] The discussion is divided into two main sections. The first describes the general characteristics of Cambridgeshire, its estate maps, the surveyors who made them, and the landowners who commissioned them. Second, a series of case studies shows how detailed analysis of individual maps enables them to be placed in their contemporary cultural and historical context.

The County

Cambridgeshire is a county on the western edge of East Anglia (figure 3.3), with two main contrasting regions. The Fens in the north are low-lying with large parishes. There were many small landowners, and manorial control was weak; fishing and fowling were important sources of income, as were fenland products such as turf and sedge; and cattle, sheep, and horses were grazed on summer pastures. Once drainage had been improved, especially after the mid-seventeenth century, the area was an important source of fodder crops such as cole- and rapeseed, onions, peas, and oats; hemp, flax, and woad were cultivated, as were improved grasses that intensified animal husbandry.[10] Retail and handicraft industries became established and diversified, stimulated by increased agricultural productivity and by land available to accommodate a growing population.[11]

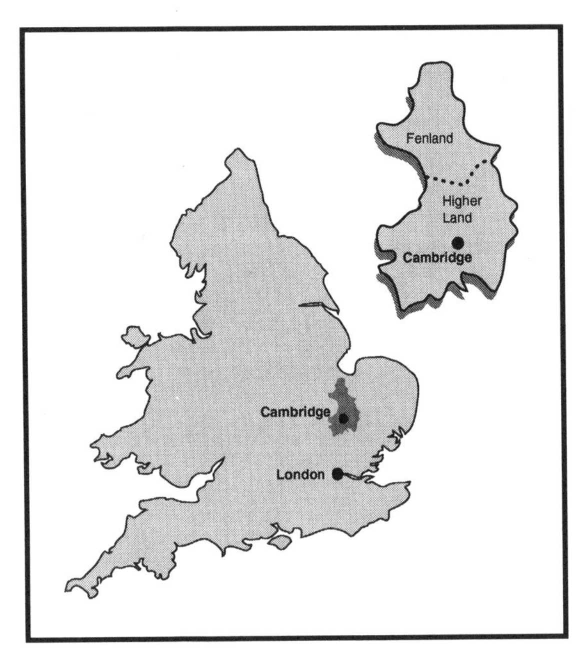

Fig. 3.3 Cambridgeshire, showing the division between the fenland and the higher ground

To the south are higher lands with smaller parishes: chalk in the center, boulder clay in the southeast, and heavy clay soils in the west. Manorial control was greater than in the Fens, and tenant farmers predominated in most parishes. It was an important area of production of wheat and barley both for London and for local markets; small acreages of fodder crops were cultivated, sheep were reared, and cattle were fattened in the western clay areas. Industries were few: stone quarrying, brick making, paper and oil mills, and basket manufacture.[12]

Two characteristics distinguish Cambridgeshire from many other English counties: the balance between private and institutional landowners, and the lateness of enclosure. Population levels rose markedly and continuously in the Fens from 1086 to 1801, but in the south fewer than half of the parishes doubled their populations between these dates, and it was not until the early nineteenth century that numbers increased. Then they did so spectacularly, more than in any other primarily agricultural county between 1801 and 1851.[13] Until the eighteenth century, however, Cambridgeshire was typical of a southeastern shire with a strong gentry community. Then, though, across the country, the aristocracy as a class failed to reproduce itself, and in southern Cambridgeshire this was coupled with a local demographic crisis. Whereas new aristocratic communities grew up in other areas of England, in Cambridgeshire only a few magnates dominated local society, notably the earls of Hardwicke and dukes of Rutland.[14] Thus Cambridgeshire had the fourth-lowest number of county families in 1860, and in 1883 was both one of the five counties with the lowest proportion of its land area occupied by properties of great landowners and, with Cumberland, the county with the lowest percentage of land owned by the gentry. By contrast, there were many small owners and, with Middlesex, Cambridgeshire was the county with the highest percentage of land occupied by public and institutional owners.[15] The church continued to be a major landowner after the Dissolution, despite the sale of much ecclesiastical land during the Commonwealth. There were a number of other institutional owners, among which the Cambridge colleges were the most significant: they owned small areas in most parishes and in 1874 accounted for 9 percent of the land in the south of the county.[16]

These landownership characteristics partly account for the other distinctive feature of Cambridgeshire: the late date at which land was enclosed. The first parliamentary act was passed in 1770, for Abington Pigotts, and it was not until 1847 that the county had almost the same proportion of land enclosed as most Midland counties. Timing of enclosure varied across the county: in general, the western clay areas were enclosed first, then the eastern boulder-clay region, and then the chalk and fen-marginal areas. In the Fens, however, much of the land had been enclosed in connection with drainage schemes, so although parliamentary acts were passed for these parishes at late dates well into the nineteenth century, only 7 percent of the land area was affected. The slowness to enclose land was partly accounted for by the large number of small owner-occupiers and absentee corporate owners; propitious conditions of a parish with one large landlord were relatively rare. In addition, there was no stimulus to convert arable to pasture land, as soil conditions favored the former, practices in open fields were flexible, and, where there was a shortage of natural pasture, common rights were vigorously defended.[17]

Therefore, estate mapping developed in a county that had two distinctive areas, little industrial development, many small and institutional owners, few large landed private owners, and late enclosure.

Cambridgeshire's Estate Maps

The database of Cambridgeshire estate maps consists of 785 items, and it is estimated that there may have been about 1,000 originally produced. These maps cover all parishes in the county except for twenty, or 14 percent. Survival partly depends on the kind of map and type of owner: sketch plans were far more likely to disappear, and indeed many probably remain unknown and unlisted in estate papers; maps that were used by later generations and so removed from their archives are likely to have been lost or to have worn out over time; and some maps must still remain unreported in private hands and so have not been found.

Over the whole period, the number of estate maps increased roughly exponentially, probably reflecting the ways in which information and ideas diffuse through society rather than any more direct demographic or economic cause. Some smaller-scale fluctuations can be tentatively associated with periods of economic difficulty or population decline, and they are borne out by developments in other counties. One of the most obvious differences is between types of owner: most maps were drawn for private owners until 1800, but institutional owners accounted for just over half of the maps overall, and indeed colleges commissioned 34 percent of the total. The last decade of the eighteenth century is particularly striking, when map production for collegiate owners increased nearly five times as rapidly as for private owners. From comparisons with six other English counties it seems that, although maps were produced at a similar exponential rate, the overall figures for Cambridgeshire were lower than those elsewhere, and mapping developed more slowly. This, too, can be partly attributed to the distinctive pattern of landownership in Cambridgeshire. Although the overall trend was upward, the use of maps increased spasmodically, and, especially in estate management, other types of document were still used, sometimes in preference to maps, throughout the period. These themes are further explored in the case studies below.

As elsewhere in the country, techniques of presentation changed on estate maps. Comparisons with developments in other counties help to show the influence of different types of landowner in determining the type of map that was commissioned, and the currency of the techniques used by the surveyor who was employed. Many maps (11 percent of the sheets) were drawn at a scale of 4 chains to the inch, 1:3,168; 10 percent of sheets were drawn at 6 chains, 1:4,752; and 8 percent at 3 chains, 1:2,376; these scales are not dissimilar from those found elsewhere. Over time, the range of scale became wider, perhaps reflecting a growth in the use of maps for specific purposes. Fewer maps, especially those drawn for institutional owners, stated their scale: many of these maps were not drawn at a specific number of chains to the inch, were little more than sketch maps, were clearly not drawn for display, and were unsigned. Fewer maps had no north point than no scale, and again maps that were drawn for private owners to display were more likely to use the latest techniques and were among the first to adopt the use of compass points. Tables were less common on maps than indications of their scale and orientation, and in general they were more frequently given on practical maps drawn for institutional owners.

Changes in decoration can similarly be interpreted in part by the owner who commissioned a map. Overall, maps of estates in Cambridgeshire broadly kept in line with contemporary styles both on other maps and in other art forms. Maps that were drawn for institutional owners were less likely to be decorated, however, whereas those that were used to show off an owner's estate and status were highly ornamented. The undecorated maps account for nearly all of those that were unsigned; surveyors were unlikely to miss an opportunity for showing their responsibility for producing works of art. Representation of buildings in plan form developed in the eighteenth century, and relatively early in Cambridgeshire. Much depended on the type of building, however: churches and the principal house were frequently shown in perspective, for display purposes, into the nineteenth century. Again, institutional owners were slower to adopt this new technique.

Planimetric and topographical accuracy of the estate maps increased over time, though not at a uniform rate, and some acreages were still estimated in the nineteenth century. Maps that were drawn for private owners to display tended to be slightly more accurate; the manor house and its surroundings could be shown in great and precise topographical detail whereas less important outlying areas of the estate were less carefully drawn. Accuracy varied according to the reason for which a map was commissioned, however, and for some the concept of planimetric accuracy was irrelevant.

Therefore the general characteristics of the estate maps of Cambridgeshire can be related to contemporary economy and society: mapmaking developed late, partly because of the particular nature of landownership in the county. Private owners, more interested in their maps as status symbols and as objects to display, were more concerned to have their maps carefully drawn and decorated and attributed to surveyors, and to show the use of current techniques, whereas in many cases institutional owners were more interested in maps as tools of estate management. Increasing numbers of maps were drawn for practical purposes, partly reflecting the growth of interest of institutional owners in their estates and of maps as useful aids in their administration, and also demonstrating a more widespread perception of the potential of cartography, by all members of society.

Land Surveyors in Cambridgeshire

The names are known of nearly two hundred surveyors who practiced in Cambridgeshire between 1600 and 1836. This figure is not unlike those for other moderate-sized English counties though, as Cambridgeshire landowners were relatively slow to have their estates mapped, they were obviously similarly laggard in employing surveyors to do so. Most of the surveyors produced only one or two maps; others drew many estate maps; and increasing numbers drew different types of map such as of roads, rivers, parks, or enclosures.

Mobility of surveyors increased in general, and growing numbers moved their businesses, though the picture is somewhat clouded by a change in type of practitioner: in the early nineteenth century, as amateurs came to draw the occasional map, many men once again operated only within the county. In the eighteenth century most men did not live in Cambridgeshire while they were surveying estates in it: many lived in neighboring counties (the example of Thomas Warren of Bury St. Edmunds has already been cited), some came from farther afield, and increasing numbers operated from London. Cambridge, too, was increasingly important: from 1785, about one-half of the estate maps were drawn by surveyors who were resident in the town, one-tenth of all those who practiced in Cambridgeshire.

The relationship between a surveyor and his employer changed: while many surveyors were dependent on patronage throughout the period, from the mid-eighteenth century increasing numbers of men worked for many employers. Indeed, about one-fifth of those who started work between 1752 and 1810 worked for more than five employers. Surveying businesses started to appear at this time, and they did more than simply measure and map land: surveying buildings and timber, and valuing land and selling it were especially popular. Others claimed interests in estate management, improving land, levelling, farming, and moneylending. Many surveyors were engaged through local recommendations, and, unless owners employed a particular man to travel the country for them, they endeavored to use men who lived near the estates to be mapped. Other surveyors advertised their services; increasing numbers did so with the rise of local newspapers in the eighteenth century.

The status of land surveyors could be quite high, and increased as their wages rose at times when general prices did not. Charges varied, however, among practitioners, areas of the country, and types of land to be mapped, and many men had sources of income other than estate surveying. Some surveyors were well educated: about one-sixth of Cambridgeshire practitioners were schoolmasters, teachers of mathematics, authors of surveying manuals, or members of learned societies. Others supplemented their income from agricultural activities, as farmers or estate stewards, and many were occupied in other land-related activities. Engineers, landscape gardeners, nurserymen, a portrait painter, an architect, a cabinetmaker, a wine merchant, and a dealer in patent medicines also mapped estates in the county.

Surveyors learned their techniques from surveying texts, through apprenticeships, and at school, and again practitioners in Cambridgeshire were not atypical. Very few, however, used more than the simplest of instruments and techniques: a chain and plane table, perhaps a circumferentor, and possibly a simple theodolite. Help from a knowledgeable local person was often essential, and surveys could be held up for considerable periods if the weather was poor.

Mapmakers in Cambridgeshire, like those elsewhere, thus enjoyed a rising status, increasingly produced a variety of maps, and became more mobile and independent. They continued to use simple techniques, usually learned through apprenticeship and at school, though with increasingly greater skill. As the ability to make maps became more widespread, greater numbers of amateurs drew a map or two, for themselves or for near neighbors. The case studies below give specific examples of many of these trends.

Landowners in Cambridgeshire

Overall, 163 private landowners and 38 institutional owners commissioned maps of their Cambridgeshire estates. At no time did a majority of the nobility and gentry have maps drawn for them, though increasing proportions did so, from about 10 percent of those who owned land in the county in 1673 to 35 percent of those in 1808 and 42 percent of those in 1860. Though more of the upper classes commissioned maps, a decreasing proportion of the whole were members of the peerage or merited inclusion in the *Dictionary of National Biography.* Increasing numbers were members of the landed gentry and clergymen. Some of these owners had county offices, and many had the typical interests of contemporary gentlemen: in antiquities, as members of the Royal Society or Society of Antiquaries, in gambling, and in sports. About one-third were educated at Oxford or Cambridge. From the mid-eighteenth century smaller landowners came to commission estate maps: an ironmonger, a carpenter, and, in the nineteenth century, farmers, brewers, bankers, a grocer, a dairyman and wheelwright, and a lime merchant. An increasing proportion of maps were of land bought by the owner or by his immediate ancestor: until 1700, nearly three-quarters of landed families had possessed their estates for many generations, while by the early nineteenth century under one-half had done so, and about one-third of owners had bought their estates themselves. This again reflects the increasing importance of smaller, local owners; so too does the growing number of people who lived in Cambridgeshire who did not own land outside the county and who only had one estate mapped. There were also two main groups of institutional owners: those connected with the university, and large London-based organizations. The latter tended to have their estates mapped earlier than did the Cambridge colleges.

Both private and institutional owners took a growing interest in their estates and increasingly paid others, tenant farmers, part-time bailiffs, and professional managers, to look after them. At the same time, however, landlords had to become well acquainted with agricultural practices to run their estates to the best advantage. Estate maps played their part in improved estate administration: nearly one-third of those of Cambridgeshire were drawn for a specific practical use. This was especially so in the early nineteenth century and for maps that were drawn for institutional owners. Maps were only one tool of estate management, however; terriers, valuations, and written surveys continued to be important and often remained more numerous than estate maps. A variety of reasons occasioned the commissioning of a map, terrier, valuation, or survey, or a combination thereof: when land was about to or had just changed hands, when a boundary was disputed, when an estate was to be let, or when land was to be improved or enclosed.

Estate maps, however, had the potential to play another role, demonstrating an owner's social standing and authority, for land was of central importance as a means for an owner to achieve status, recognition, and influence among his peers. Thus many maps were drawn for display and were beautifully decorated, sometimes including the owner's coat of arms; they may be on rollers, hanging in front of a bookcase, or in a binding suitable for library shelves. Surveying texts gave instructions on how such maps should be produced, and there is every indication that the details were carefully followed. Institutional owners, too, commissioned such maps, for they also had the desire to impress, but private owners account for most of these maps.

Most of the private owners in Cambridgeshire employed local men to work for them, so, too, did many of the Cambridge colleges. Indeed, a few local practitioners in the late eighteenth and early nineteenth centuries must have worked more or less full-time for the colleges, many of which tended to employ one main surveyor at any one time. These men were more than estate mappers: they valued land, carried out surveys, managed woods, collected rents, supervised repairs, and found suitable tenants.

Cambridgeshire, therefore, had distinctive characteristics as a county. By studying a series of maps in great detail and relating them to both local and national social and economic developments, it is possible to assess the roles the plans played in contemporary society.

Maps, Makers, and Owners

Five sets of estate maps are analyzed here in detail, to show why they were drawn, how they were used, who the landowners were, what was their awareness of the potential value of maps, who the surveyors were, how important land surveying was to them, and where the maps fit among contemporary noncartographic documents. These examples have been chosen to show the variety of maps, owners, and mapmakers who were present in early modern Cambridgeshire.

Maps of Newly Acquired Estates: The Corporation of the Sons of the Clergy

In 1678, the Corporation of the Sons of the Clergy was established by charter to assist the widows and children of deceased clergymen. From a series of legacies, the corporation became a substantial landed proprietor with estates in Berkshire, Cambridgeshire, Kent, Northamptonshire, and the East Riding of Yorkshire, yielding an annual income of nearly £3,000 in the early 1720s.[19] By 1716 the corporation had acquired about 235 acres in the Fens, in the parishes of Wisbech St. Mary, Tydd St. Giles, Upwell (extending into Norfolk), and Tydd St. Mary (Lincolnshire). At about this time, too, 187 acres were bought in West Wratting Parish in the southeast from Tyrell Dalton, and 192 acres in Willingham Parish in accordance with the will of Samuel Saywell.[20]

These estates were mapped, shortly after their acquisition, by Arthur Frogley about whom little is known.[21] He was employed by the corporation between 1714 and 1719 to map their land in Essex and in Cambridgeshire, and in 1724 he mapped land in London belonging to the duchy of Cornwall.[22] The map of the West Wratting and Willingham estates (figure 3.4) is drawn on parchment at just over 4 chains to the inch, is 35 inches high and 30 inches wide, and is not particularly well finished. It shows buildings in perspective, individual trees, roads and gates, and watercourses. Color is used to distinguish the land belonging to the different farms; the title is in a cartouche with classical columns and a pediment, fashionable for the time; and orientation is shown by a compass rose. A table within the cartouche gives tenants, field names, and acreages. Here, then, is an example of an institutional owner having its estates mapped so that it can see the land it has acquired and the tenants of

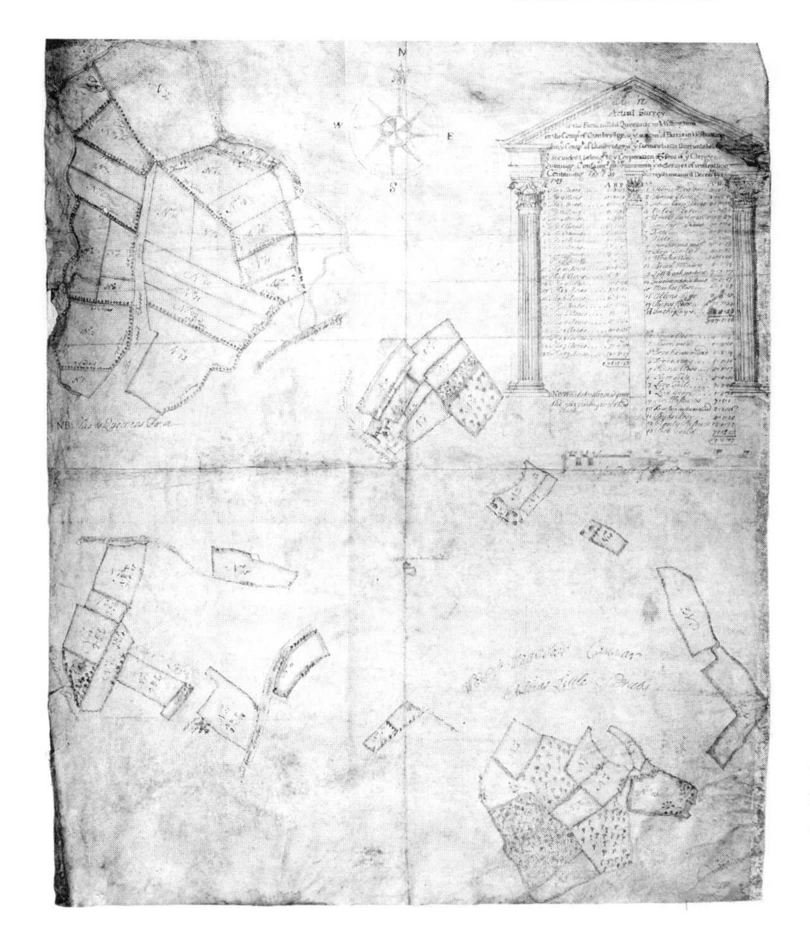

Fig. 3.4 Arthur Frogley, West Wratting and Willingham, 1719 (By permission of the Syndics of Cambridge University Library, Map Room, MS Plans 716)

each plot. The maps are functional, not by a well-known surveyor, and not elaborately decorated.

The corporation continued to take an interest in its lands and realized that careful management of its property by improving estates, negotiating new leases, visiting tenants, and calling in arrears of rent and tithes was necessary to generate the income to be distributed to the needy.[23] The West Wratting estate was remapped in 1737 by John Bowles, a more established surveyor than Frogley and a map seller too,[24] and a superior product was the result, with the title in a gold baroque cartouche.[25] The other Cambridgeshire estates were not resurveyed until the nineteenth century; presumably the existing maps were considered to be adequate for estate management until then.

From the mid-eighteenth century the estates were inspected frequently as a means of improving control over the tenants, and in the early nineteenth century, too, repeated visits were made. Property in Cambridgeshire was inspected in November 1809 and again by the treasurer, John Bacon, in 1811.[26] At about this time, Frogley's maps of the estates were copied by John Newton of

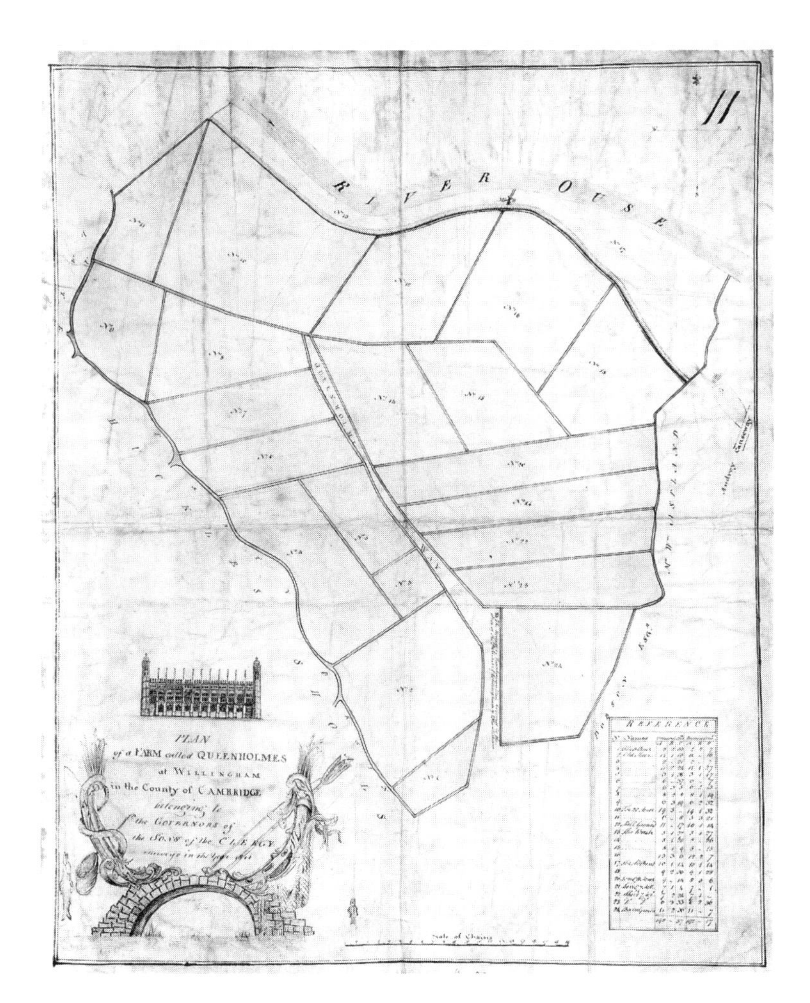

Fig. 3.5 Anonymous, Willingham, 1811 (By permission of the Syndics of Cambridge University Library, Map Room, MS Plans 717)

Chancery Lane, and in 1811 the estates at Willingham and Wisbech were resurveyed and mapped.[27] The Willingham map (figure 3.5) does not show substantial changes from the position one hundred years before (Frogley's plan of Willingham is shown at the top left of figure 3.4), though additional drainage ditches have been dug; no trees are shown, but it is more carefully drawn and has a vignette of the chapel of King's College, Cambridge. It is unsigned, which is unusual for a map that is drawn with some style. The estate at West Wratting was not mapped in 1811, probably because it was undergoing enclosure at that time, but, on completion, a map of the newly allotted estate was taken from the map that accompanied the enclosure award (figure 3.6).[28] This map is just one example from the many Cambridgeshire estate maps that were drawn to show land allotted on enclosure. Indeed, parliamentary enclosure was probably the most influential stimulus to production of estate maps in the early nineteenth century: in the first quarter of the century, 45 percent of the par-

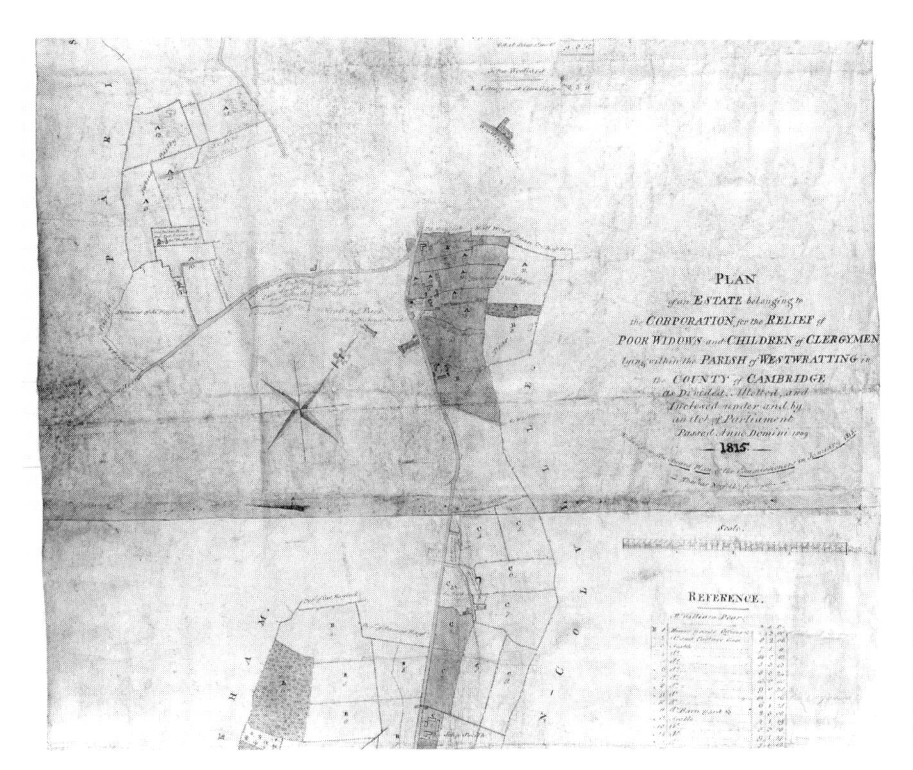

Fig. 3.6 Thomas Norfolk, West Wratting, 1815 (By permission of the Syndics of Cambridge University Library, Map Room, MS Plans 714)

ishes were enclosed, and 86 percent of the maps that were drawn were of these parishes.[28]

The Corporation of the Sons of the Clergy, therefore, demonstrates an institutional owner interested in its estates and their management: land was mapped on its acquisition and then about a century later in connection with inspection by officials from the organization and with enclosure. The maps that were produced were not particularly splendid; they were rarely by particularly well-known surveyors and were sometimes unsigned. As the next example shows, other documents must also have been used in the management of the estates, and the maps were just one of the available tools.

Maps and the Renewal of Leases: Queens' College, Cambridge

Queens' was one of the first Cambridge colleges to commission maps of its estates: on 3 July 1764 it was "agreed that the Master may order a map to be made of our Estate at Capel in Suffolk," and on 10 January in the following year he was "empowered to order our Estate at Bumstead [Helion Bumpstead, Essex] to be survey*e*d and mapp*e*d if he shall judge proper."[29] So began a tradition of mapmaking unparalleled elsewhere in the university.

To discover the reason for these maps, one has to delve a little further. In the Register of Leases, it is recorded that the recommended fine for renewal of the lease of Capel in 1765 was £131 12s.

7d. for a 258-acre estate, but as the fine had previously been only £80, it was agreed to limit the new one to £100. Soon after the fine had been set, however, "a Map of this Estate (which had been order*ed* the Summer before) was brought in, and that makes the number of Acres by mea*sureme*nt 298A 2R 36 Po*les* exclusive of the House, <&> Homestall & Hedges. Regard therefore should be had to this at the next renewal of the Lease."[30]

Some explanation of the system of financing colleges and their activities is necessary. Their endowment was in land, which was usually let out to tenants. In 1571 and 1572, legislation in Parliament limited the length of leases of agricultural property to three lives or twenty-one years, and of urban property to forty years, and an act in 1576 stipulated that colleges had to draw one-third of their rent in kind.[31] As prices of wheat and malt rose, so did collegiate incomes. This increase was to be applied to the commons of the college, and surpluses were divided among those who were entitled to dine on the foundation.[32] At the same time, rents tended to remain level, and chief gains from growing agricultural productivity were realized from fines, charged for the renewal of a lease every seven years, which were related to the real value of the holding. These fines were divided among the master and fellows as their "dividend." Thus interest centered on collecting rents and entry fines, which were insensitive to market fluctuations. To make an attempt to reflect current prices, it was necessary to have the estate surveyed each time a lease was renewed; this could be done, but did not become common until the 1770s, presumably as a result of rising land values.[33] It was not until this system of letting land on beneficial leases had been replaced by letting land at rack rent, however, that colleges could hope to derive a realistic income from their estates and were thus encouraged to become more than mere rent collectors.

In 1818 the advantages of letting at rack rent were set out for the fellows of Queens' College.

> It is well known that College Estates are usually let on what are called beneficial Leases, renewable at any time, upon paying Fines corresponding to the No. of years lapsed; but they are in fact renewed most commonly every seven years.
>
> This Practice is founded on the mutual advantage of both Parties. The Lessees are hereby encouraged to cultivate & improve the property on the expecation [*sic*], of paying a reasonable Fine from time to time; & the members of the College obtain a due share of the Profits, while they are at the same time benefiting their successors.
>
> It happens, however, frequently, that the aforesaid process of renewing leases is stopped on account of the Lessees, being either unable or unwilling to pay the requisite Fines, even in cases where the Fines are stated much below their real value, according to the Judgment of surveyors of experience & fair character: & further, in some instances, the Leases have actually run out, & a purchaser of a new Lease could not be found to pay a very reasonable demand made by the Master & fellows.
>
> Under such circumstances, the Mr & fellows, in certain instances have ventured to borrow Money . . . & divided the said Money so borrowed <would have amounted to> among themselves, just as if the Fine had been actually paid & the Leases renewed.—And, as a part of this process, the Property has

> been let at a Rack-rent, & the leases suffered to run out; . . . & where this has been done with discretion & a proper regard paid to the several Interests . . . the College estates have been much improved & rendered more productive both to the present & future Society.[34]

Queens' was a poor college that rarely made a substantial surplus and was frequently in deficit. It did, perhaps as a consequence, devote a higher percentage of its expenditure than some of the wealthier colleges to surveying its estates, and it kept up levels of spending even after financially unsuccessful years.[35]

The college's use of maps can be seen from surveys of its estate at Coveney in Cambridgeshire. At the audit of 1784, it was reported that, "[o]n the approaching expiration of the 3 Leases of this little Estate, a Surveyor was sent to value the Land, and to Map the 2 Closes Call*ed* Hall Closes and the Meadow call*ed* New Meadow."[36] Joseph Freeman had been employed and was paid £2 12s. 6d. for his map and valuation (figure 3.7).[37] The map is a working document, drawn on paper, with a table giving field names, acreages, land use, and value. The table includes a mistake: the acreage of plot 3, 1 acre, 2 roods, 23 perches, has been mistranscribed as "1a.2r.33p." Beneath the table is a note that makes the purpose of the survey quite explicit: "Observations. These Lands appear to have been neglected they are very Poor but may be greatly Improved by Draining and Manure; if a Lease of 21 Years was granted I would advise there being let as here valued for the first seven Years to encourage the Tenant to Improve, the rest of the Term at 20 Shillings an Acre, viz £21 4S 6D at the expiration of Seven Years. Valued by Joseph Freeman in the Month of May 1783. Fen Land in the Occupation of Waterlow 7S p*er* acre." This advice was heeded. When the lease was next renewed in 1806, the estate was surveyed but not mapped: on 27 August 1805, Joseph Truslove charged six guineas for three days' work: "Journey to Coveney to look over and value 2 farms there; from thence to Willingham Flatt, to look over and value 2 farms there and esti-

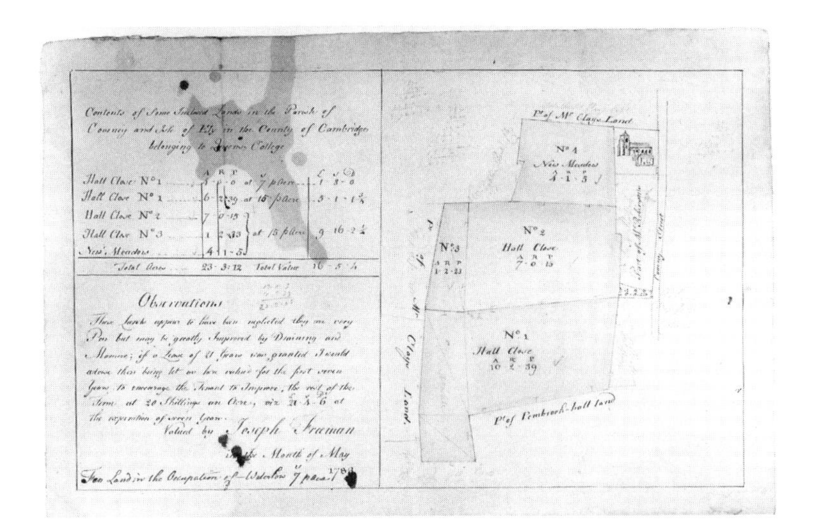

Fig. 3.7 Joseph Freeman, map and valuation of Coveney, 1783 (Queens' College, Cambridge, 328)

mate the dilapidations of Bath's Cottage, Barn, and Stable; from thence to Willingham to view and value a Farm in the occupation of Wayman," and £2 17s. for his expenses. A further five guineas were charged for, "[m]aking the Calculations and Valuation of said five farms or parcels of land, with fair Copy thereof and Observations subjoined, as to the mode of Cultivation & repairs of Buildings; also attendance therewith at College."[38]

The estate was valued and mapped again in 1817 (figure 3.8), Alexander Watford was employed this time, and again functional maps were produced, undecorated, on valuations. Comparison with Freeman's map confirms the impression that Watford's plans were merely sketches to show the layout of the land: the lower sketch shows how Hall Close, number 1 in the earlier map, has been subdivided into plots 5, 6, and 7, let to Richard Clay. The most useful part of the document for the college are the observations, which state that there were still trees on the property though it was not thought necessary to include them on the sketches.

> The Lands in Lease to Messrs Clay, are a very compact and valuable little property. I did not find the Land held by Mr Rich*a*rd Clay, in such good condition as that of Mr Char*le*s Clays, but I have not made any allowance for this, and have valued the Land on the same Scale. No 7 from situation and nature is much worse, and I should feel no hesitation in recommending this to be ploughed for 3 Years, and after that time laid down again. There are some very handsome Trees in the Hedgerows, but as there are none but what will improve, and as they are very Ornamental to the Country, I should not take them down until there was appearance of their getting worse by standing.

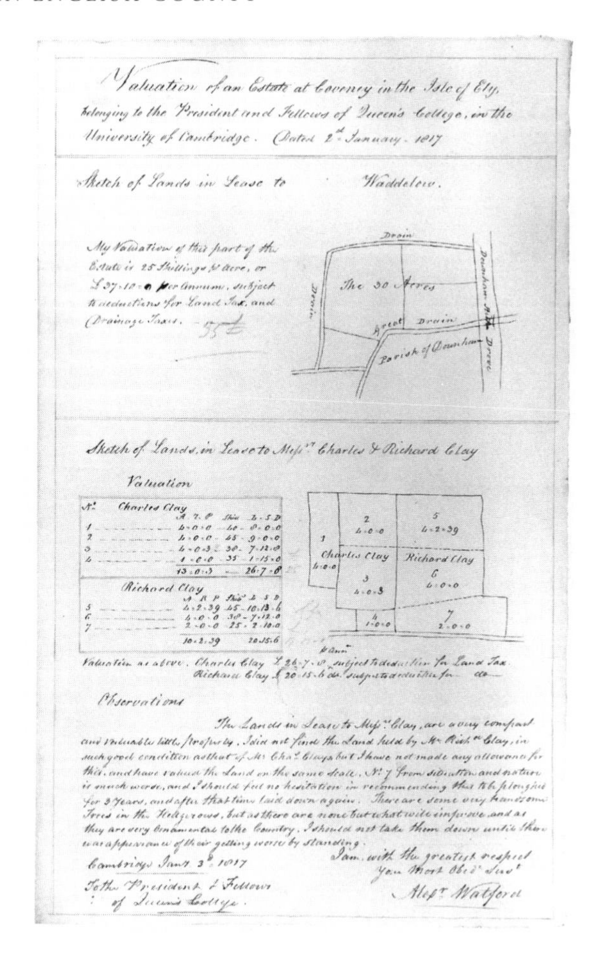

Valuation of an Estate at Coveney in the Isle of Ely, belonging to the President and Fellows of Queen's College, in the University of Cambridge. Dated 2d January 1817

Sketch of Lands in Lease to Waddelow.

Sketch of Lands, in Lease to Messrs Charles & Richard Clay

Valuation

No	Charles Clay	A R P	Shils	£ s D
1		4-0-0	40	8-0-0
2		4-0-0	45	9-0-0
3		4-0-3	38	7-12-0
4		1-0-0	35	1-15-0
		13-0-3		26-7-0
	Richard Clay	A R P	Shils	£ s D
5		4-2-39	45	10-13-6
6		4-0-0	38	7-12-0
7		2-0-0	25	2-10-0
		10-2-39		20-15-6

Valuation as above. Charles Clay £26-7-0 subject to deduction for Land Tax.
Richard Clay £20-15-6 do. subject to deduction for do

Observations

The Lands in Lease to Messrs Clay, are a very compact and valuable little property. I did not find the Land held by Mr Richd Clay, in such good condition as that of Mr Chas Clays but I have not made any allowance for this, and have valued the Land on the same Scale. No 7 from situation and nature is much worse, and I should feel no hesitation in recommending this to be ploughed for 3 Years, and after that time laid down again. There are some very handsome Trees in the Hedgerows, but as there are none but what will improve and as they are very Ornamental to the Country, I should not take them down until there was appearance of their getting worse by standing.

Cambridge Jany 3d 1817

To the President & Fellows of Queens College.

I am, with the greatest respect
Your Most Obedt Servt
Alexr Watford

Fig. 3.8 Alexander Watford, valuation and maps of Coveney, 1817 (Queens' College, Cambridge, box 20)

Two-thirds of the maps made for Queens' College were for a similar purpose, although once again it must be emphasized that maps were reused and were only one of a range of documents that were deployed. For example, Joseph Truslove was involved in a dispute between the college and its tenant at Shudy Camps in Cambridgeshire before a lease was renewed in 1811. He visited the land on 18 October; then he spent three-and-a-half days from 12 November "examining documents with the Master as to the Lands in dispute, procuring old Plan of Shudy Camps, scaling off the pieces and parcels of Lands to ascertain the quantity comparing the documents with the Plan—Making Plans for Coll*ege* and Mr Bridge [the tenant], drawing out a Case for Mr Bridge of the discription of Lands in dispute with the abuttals & boundarys, Copy of Coll*ege* Terriers, Valuation and calculations of the Fine with proposals of the Terms and conditions of Letting him the Farm—And also making fair Copies of Plan, Case Statement &c for Mr Bridge."[39]

In 1822, however, the college decided to commission an atlas of all its estates: "Agreed that Mr Watford do value for those Estates the leases \of which/ come on for renewal, which have not been valued in the course of the last fourteen years; and that Plans be made of all our Estates on the Same Scale."[40] This was no mean undertaking: the work took three years for Alexander Watford and his nephew, James Richardson, to carry out, and the college could barely afford it. The bill was queried, inspected by Thomas James Tatham, surveyor of Bedford Place, London, and a reduction was recommended by him.[41] Even so, partial settlement of the account in 1828 (£457 of a total bill of £757) was responsible for over 21 percent of the total expenditure of the college in that year.[42] It is hardly surprising that Trinity College decided that such an atlas would be too expensive for them to commission: the senior bursar noted on 15 December 1826: "Mr Watford's estimate for new plans of our southern estates. Having this before me, I dare not advise the board to employ him—much as it seems to me desirable to have such plans in our possession."[43]

Two copies of the atlas were produced: one on parchment for the master, with a decorated title page and the coat of arms of the first founder, Queen Margaret of Anjou, and an alternative seal for the college facing it; and an inferior paper copy for the bursar. Figure 3.9 shows the Conveney estate in the master's copy: it shows buildings, state of cultivation, acreages of plots, hedges and fences, and roads and gates, and has a table giving tenants, field names, acreages, and land use. It shows a much closer resemblance to Freeman's map than to Watford's rough sketch; it is interesting to note, however, that Freeman's drawing of Coveney Church is more accurate than that in Watford's carefully drawn plan for the atlas.[44]

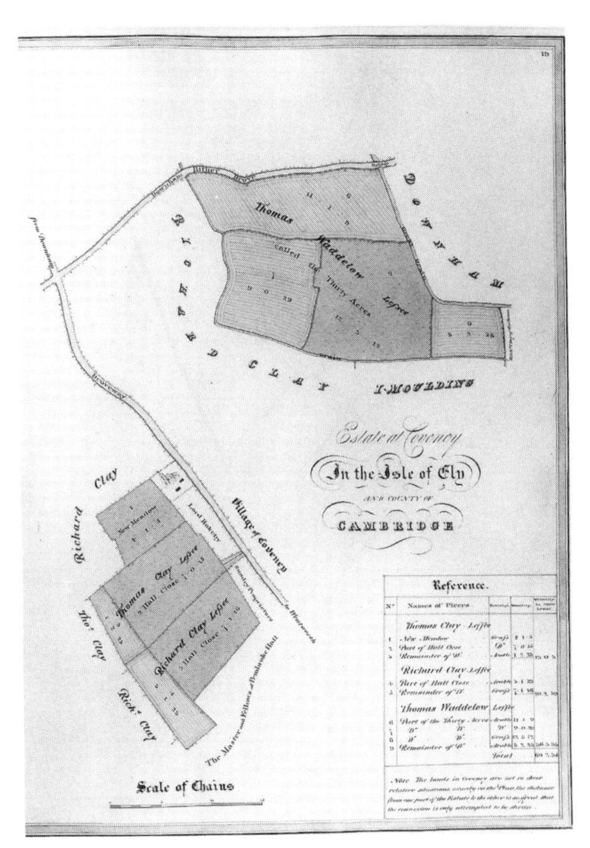

Fig. 3.9 Alexander Watford, Coveney, 1825 (Queens' College, Cambridge, 355 B19)

The atlas was to be both a permanent record of the college's estates and something to be proud of. It was kept up-to-date for a few years: the Abbotsley estate in Huntingdonshire was enclosed in 1838, and in 1839 a bookbinder removed from the master's copy the plan that had previously been made, so that it could be replaced with one by Charles Day.[45] Unfortunately all did not go smoothly: Watford wrote to Day on 18 June 1840 that "[t]he Plan you have sent is so very different from the Size of the Master's Atlas (by some means or other) that it cannot be committed there. The Bursar thinks that 12 Guineas under the Circumstances will be a handsome remuneration to you as he must have it in the <rough> Bursars rough Atlas." (Day's bill had been for £157s.). In addition, Watford wrote to the bursar, "Will you be kind enough to let me have by your Confidential Servant the Masters set of Plans and your Own and I will see what can be done with Day's plan."[46] The situation could not be saved, however, and the Abbotsley plan is to be found in the bursar's copy today.

Queens' employed forty-eight surveyors between 1600 and 1836, and forty of them lived near the estate to be mapped. Fewer than half, however, drew maps, and similar figures can be found for other colleges. One main surveyor was employed at any one time: Joseph Freeman worked for the college between 1768 and 1787, and Alexander Watford was the principal surveyor between 1821 and 1839. Neither of these two men were of national importance, but both were significant local figures who worked for many colleges and private owners, and drew over fifty estate maps apiece.

Joseph Freeman was born in 1734 or 1735 and moved to Cambridge from London in 1764.[47] His work for at least eight Cambridge colleges, the university, and six private landowners took him to fourteen different counties, mostly in eastern England. He made maps; acted as a land agent, tithe surveyor, valuer, and auctioneer; and was steward to St. Thomas' Hospital and the Charterhouse in London. A number of men connected with him may have been assistants: James Backer in 1766; Nicholas King; who was also a draftsman, in 1791–92; and Thomas Watson in 1794–97. Freeman had an apprentice, Francis Marshall, possibly a relative by marriage, who set up on his own in 1792.[48] A shy and diffident man, Freeman has a secondary occupation as a copier of portraits and again was employed by many Cambridge colleges in this role; his brother, Charles, painted buildings.[49] In the 1790s, however, land surveying and related work were practically Freeman's sole activity. He earned over £2,000 from land surveying and mapping in the course of his career, left goods worth over £600 on his death in 1799, and was sufficiently important to have a brief obituary notice in the *Gentleman's Magazine*.[50]

Alexander Watford was the son of a surveyor, also called Alexander, who was employed by Queens' College from 1792 until his death in 1801. Shortly after the elder Watford had started working for Cambridgeshire owners, the family moved to Cambridge from Bedford. The father mapped estates, was engaged on enclosure work, both as a surveyor and commissioner, and acted as a land agent in twenty-two counties from the North Riding of Yorkshire in the north to Sussex in the south and from Carmarthenshire in Wales in the west to Suffolk and Essex in the east.[51] For Queens' he, too, mapped estates before leases were renewed; this was not always a trouble-free process. On 10 February 1801, two months before his death, he surveyed land at Eversden in Cambridgeshire and "had infinite trouble on account of the badness of weather & shortness of days, being obliged to proceed on acc*oun*t of settling Fine," and charged one shilling per acre.[52] His estate was valued at £1,000 at probate, more than Freeman's.[53]

The younger Watford was first employed by Queens' in December 1801, when he completed his father's survey of Eversden with the aid of the tenant and a local person who showed him the boundaries: "4 days going over to Eversden, finishing the Survey, taking a discription of every piece of Land, making a terrier and correcting the same with Mr. Rycroft and the person that went round the Bounds with me . . . 2 days making a Fair Terrier, making the Old & new one coincide, and correcting Maps."[54] However, he carried out little work for the college until 1821; from then on he was much employed and was paid over £1,885 by it, more than he earned from each of a number of other colleges. His work took him as far afield as his father, and he became involved in a number of land-related activities, as an enclosure surveyor and commissioner, a valuer, an attorney, an auctioneer, a property agent, and a timber valuer.[55] He was responsible for at least 125 maps between 1801 and his death in 1844, and his office produced about the same number of valuations, and also terriers and written surveys. He employed assistants and local men to show him the area and to carry his chain. Watford achieved a considerable status: he arbitrated in local disputes, was an overseer for the poor, dined with local academics, had an interest in antiquities, corresponded with national figures (in 1842 he presented a verse rendition of the collects of the Church of England to the archbishop of Canterbury, who declared the possibility of is publication "hopeless"), left an estate valued at £2,000 when he died on 17 June 1844, and was described as an "eminent surveyor" in the *Gentleman's Magazine*.[56]

These three maps of Coveney, therefore, provide an entré into the world of collegiate estate management in the late eighteenth and early nineteenth centuries, and to two of Cambridge's most influential families of land surveyors. Maps were one of the tools of estate management, but only one; they were one of the products of land surveyors, but only one, and must be seen in this context. Ideas and information spread between colleges, and thus estate maps and atlases could be more than merely functional tools; if elegantly produced, they could also be a means of impressing one's colleagues. Similarly, surveyors worked for a number of the colleges, and even the best established of them used rudimentary techniques and depended on local knowledge and favorable crop and weather conditions. The following example enables comparison with another collegiate landowner.

Maps and the Exchange of Property: Emmanuel College, Cambridge

Although Emmanuel College had its land at Threadneedle Street in London shown on a lease in 1703,[57] this early promise was not realized in a wealth of cartographic documents to be found in the archive today. Indeed, the only map of its Cambridgeshire property is a sketch of an area of less than 1 acre, drawn in connection with a proposed exchange of land (figure 3.10).

The map of Eltisley is drawn to show land in the parish along the boundary with Papworth Everard, for which there was a suggestion that it might exchanged with land owned by Charles Madryll Cheere. It is a plain map; all it shows is the land and its quantity at a scale of just under 3 chains to the inch, four plain compass points, and a table summarizing the acreages. The parish boundary is colored red; the rest of the map is uncolored. There

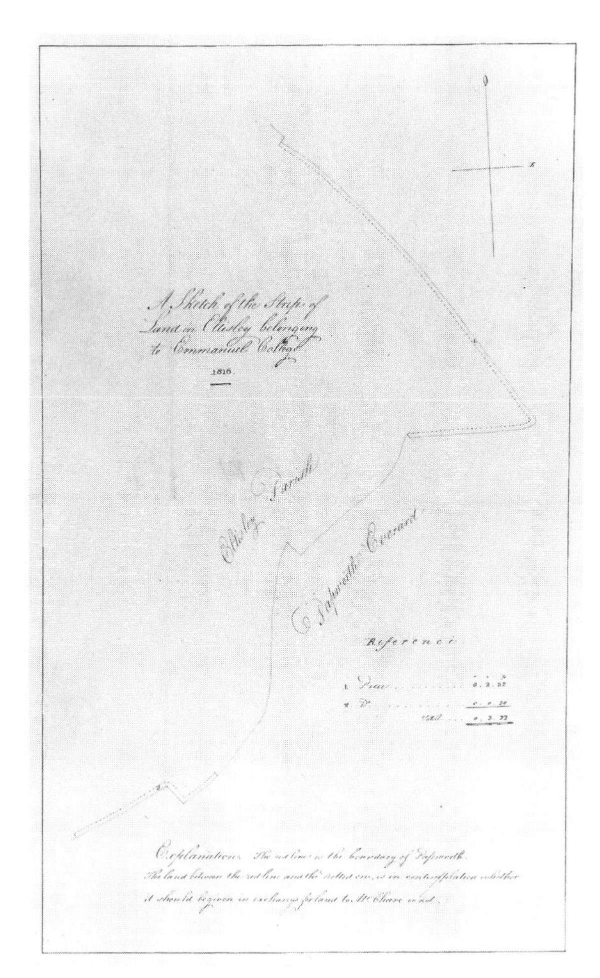

Fig. 3.10 Anonymous, Eltisley, 1816 (Emmanuel College, Cambridge, box 12.C17)

is no scale, and the sketch is not signed. This is another example of a map that was drawn for a specific practical purpose.

The absence of maps in the archive, however, should not be taken as an indication that the college was uninterested in the estate. From the mid-eighteenth century a number of visits to it are recorded. In 1745 the bursar inspected the estate, and in the following year three shillings were spent on having it surveyed.[58] After this survey the estate was let for twenty-one years, but no fine was levied, as the tenant was to improve the land.[59] The estate was visited again in 1754, on 16 and 18 October 1777 Joseph Freeman surveyed and valued the land and compared terriers of it, and he returned again on 3 June 1778.[60] William Custance was paid for selling timber from the property in 1802 and 1803, [61] and he was concerned with letting the estate between 1812 and 1814. He inspected repairs that had been carried out on 27 and 28 November 1812, valued the farm the following April, and then advertised its availability for lease.[62] In 1816 there was the possibility that a small part of the estate might be exchanged.

This is an example of a college that had relatively few maps drawn; it did, however, have its estates surveyed, often in connection with the renewal of leases, and local Cambridge practitioners, Joseph Freeman and later the Watfords, were employed. The college also used a firm of London agents, Messrs. Nettleshipp, to engage men on its behalf, and asked them to recommend as surveyors men who lived near the estates to be surveyed.[63] The map of Eltisley was drawn to show land along a boundary with that of another owner, and has affinities with one of the maps in the next set to be considered.

Maps for Business and Pleasure: Benjamin Keene of Linton

Catley Park in Linton had a succession of owners in the mid-eighteenth century: it was sold in 1764 to Thomas Bromley, Lord Montfort, to pay the debts of Thomas Sclater King. In his turn, Monfort sold it in 1772 to the bishop of Ely, Edmund

Keene, who shortly afterwards transferred it to his son, Benjamin.[64] Benjamin had the land mapped twice.

The first map (figure 3.11) is another example of a map that was drawn to show a boundary, for which the responsibility for maintenance was in dispute. Benjamin Keene and his neighbor Thomas Wolfe disagreed over who had to make and maintain a hedge, ditch, and fence between Catley Park and Burton Wood when an existing fence was removed, and a plan accompanies Thomas Pennystone's judgment of the case. Only the disputed area is shown, the type and length of the boundary, and its immediate surroundings, at nearly 2 chains to the inch. On the plan is a written explanation of the boundary: "The Red dotted Line denotes the course of the Ditch between Burton Wood and Catley Park Grounds,—From A. the begin*ning* of Burton Wood (at the end of the old Lane by Hadstock Field) to B. the corner of Catley Park Grounds which is 6 1/4 Rods Mr Wolfes Hedge and from thence to C. between *sai*d Wood & Catley Park G*roun*d in extent 72 Rods is the Hedge in dispute; from thence forw*ar*d to D. the corner of Catley Garden Wall is 7 1/4 Rods Mr Wolfes Hedge & from thence forw*ar*d to E. being 26 1/2 Rods more Mr Wolfe likewise acknowledges to be his Fence." The judgment, however, does not refer to the plan:

> I do hereby declare my full opinion and Judgment. . .That the Hedge, Ditch and Fence between Catley Grounds and Burton Wood,

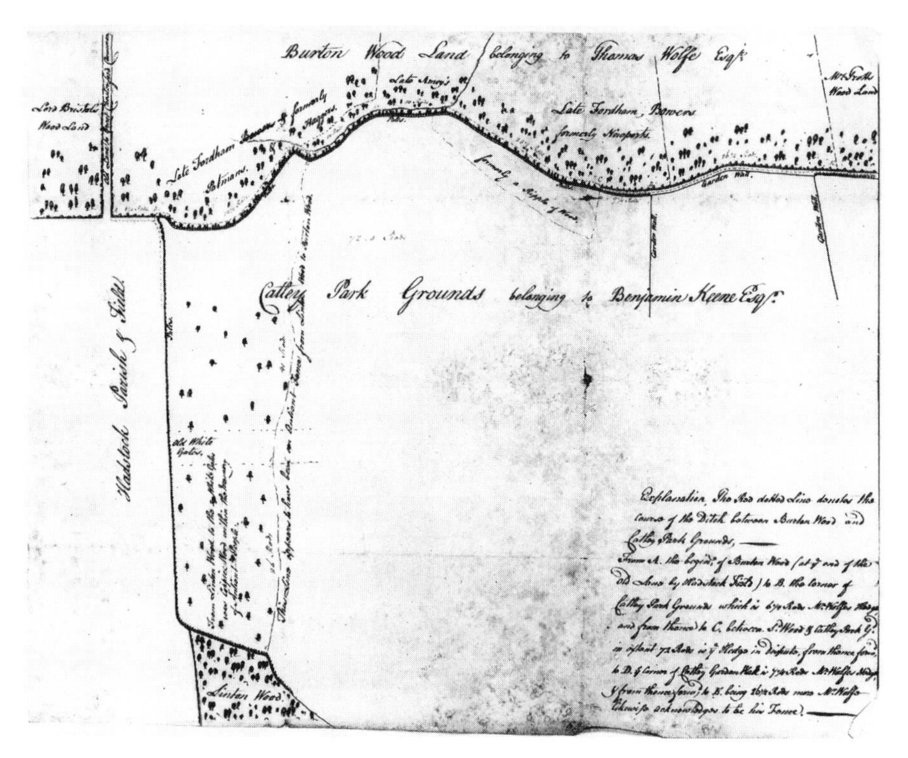

Fig. 3.11 Anonymous, boundary to Catley Park, Linton, 1777 (County Record Office, Cambridge, R59/14/11/11e)

beginning from the South East Corner of Catley Grounds (next to Hadstock) and going North Westward, to the Turn of the Ditch and Bank within Seven Rods and A Quarter, or thereabouts, of the Corner of the Brick Wall of Catley Garden for the Length of <> Seventy two Rods, or thereabouts, <> Did and doth now of Right belong to the owner of Catley Ground to make maintain and uphold

And that the Hedge Ditch and Fence from such turn of the Ditch and Bank to the End and Extent of the said <> Mr. Wolfe's property in the said Wood, North Westward, <> did and now doth of Right belong to the owner of the Wood, to make maintain and uphold, to the best of my knowledge Judgment and belief, as also upon the best evidence of the Antient Men before mentioned.

This map, therefore, does not form an integral part of the adjudication; it does not have a legal status, but was drawn to illustrate a written document and to show precisely which areas was being described.

Two years later, in 1779, Keene commissioned an entirely different type of estate map: a plan of the land by a surveyor who has already been mentioned several times, Joseph Freeman. The map (figure 3.12) is large, 70 inches high and 30 inches wide, is drawn on parchment at about 3 chains to the inch, and is decorated. It shows part of Keene's Catley estate, with perspective drawings of Catley Park, Little Linton Farm and its buildings, Hilders-

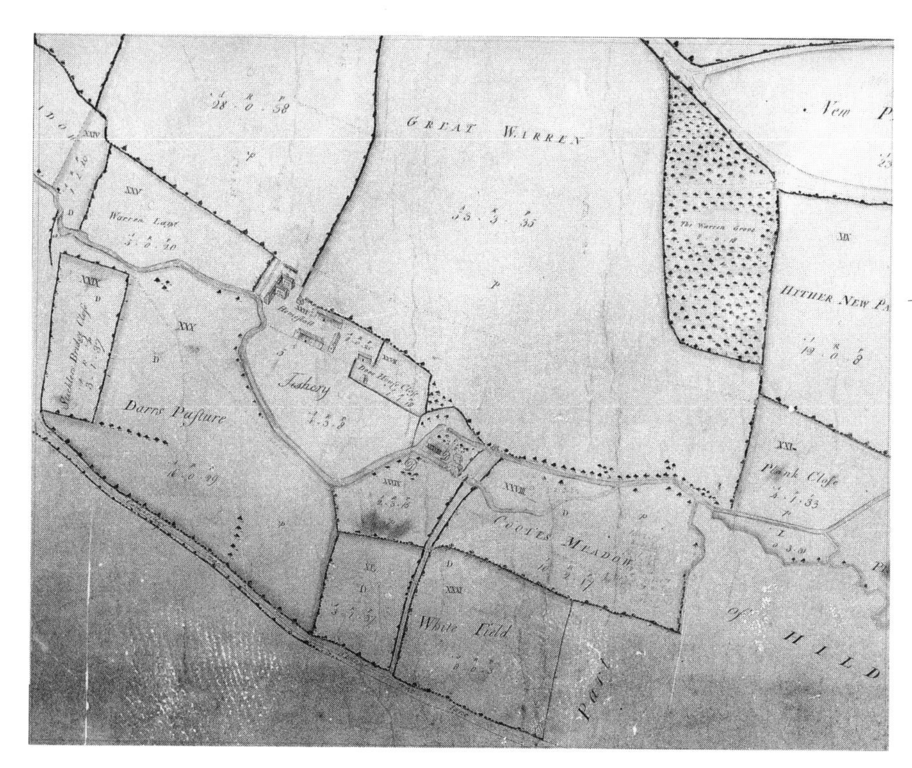

Fig. 3.12 Joseph Freeman, part of map of Catley Park, Linton, 1779 (County Record Office, Cambridge, 124/P64)

ham and Little Linton mills, and a dovecote. Figure 3.13, an aerial photograph of Little Linton Farm that was taken in 1961, shows that Freeman's drawing of the farmhouse gave accurate architectural detail, and there is every reason to believe that the other buildings were also true representations. Woods are drawn and trees, meadowland, pasture, fisheries, moats, bridges, gates, hedges, fences: a wealth of detailed topographical information. The title has a rococo cartouche; there are a scale and compass points. A table on the map gives tenants, field names, and acreages.

These details are in contrast with Freeman's other map shown here (figure 3.7). Examples could have been given, however, of the beautifully decorated maps that he drew for Queens' College of its

Fig. 3.13 Aerial photograph of Little Linton Farm, 1961 (Cambridge University Collection of Air Photographs: copyright reserved)

Fig. 3.14 Monument to Mary and Benjamin Keene in Bartlow Church (author's photograph)

estates in south Cambridgeshire, Suffolk, and Essex at about this time, which similarly show more important buildings in perspective view and, significant for a college that set much store by its timber resources, the number and species of trees in the hedgerows. Or Freeman's later maps of the 1790s could have been illustrated, which were drawn in a much plainer and more restrained style with simple coloring. Those maps that were drawn for colleges such as Clare and St. John's were decorated in such a way; so, too, were the maps of West Wratting drawn for the duke of Bedford in 1794.[65] Here, once again, display was important, and so the duke's coat of arms was added, magnificently colored, to what were otherwise relatively subdued maps.

Benjamin Keene was born in 1753, educated at Eton, and admitted as a fellow-commoner to Peterhouse, Cambridge, the college of which his father had been master, in 1770. He became member of Parliament for Cambridgeshire in 1776 and was high sheriff in 1804.[66] In 1806 he leased Westoe Lodge in Castle Camps and made it his main seat; he purchases land there from the Charterhouse in 1810 and immediately had it mapped;[67] and he had bought the freehold of the lodge by 1825.[68] He added to his estate in Linton in 1783;[69] he did not commission a map of it, and perhaps he owned and was satisfied with the existing survey and map of 1732.[70] He was an educated, landed gentleman who held county office, endowed a school,[71] and was commemorated on his death in 1837 by a monument in Bartlow Church (figure 3.14).

Catley Park was mapped to show Benjamin Keene his land soon after he acquired it, no doubt for practical reasons so that he knew the extent of his property and could administer it accordingly. This was not a map to be hidden away with other estate papers, however; it was to be seen and enjoyed, and displayed to others as evidence of Keene's territory and his ability to employ a well-known local surveyor to map it.

These two maps therefore show contrasting items drawn for a typical member of the landed gentry: a map used to settle a boundary dispute, and a map drawn as a status symbol. A member of the aristocracy similarly commissioned two types of map, and these will be discussed in the final case study.

Maps for Estate Development and Record: Edward, Lord Harley

John Holles, duke of Newcastle, was a member of a family that took great interest in its estates. He saw county affairs in terms of estates, patronage, and personality, and managed his land himself, demonstrating a deep interest and understanding of estate management. From 1700, once he had brought order to property that he had inherited in poor condition, he expanded his holdings: for political reasons he bought control of two parliamentary seats, and then he combed the east Midlands for large estates that offered a good return for a purchaser who was sufficiently wealthy to buy them complete. Among these new acquisitions, he bought the Wimpole estate in Cambridgeshire from the earl of Radnor. As a demonstration of his interest in land management, he gave his land agents a list of thirty-one questions to ask when they were considering an estate for possible purchase. These questions covered every aspect of an estate's value, natural resources, taxes and charges, conditions of the tenants' houses, leases and arrears, and the capacity of the land for improvement and enclosure.[72]

Holles's daughter Henrietta inherited her father's property at Wimpole, and she married Edward Harley in 1713.[73] Harley was little interested in estate management or politics; he was far more concerned with landscape gardening, building, and collecting books, manuscripts, pictures, medals, and miscellaneous curiosities. This is the owner who was responsible for two sets of estate maps of Cambridgeshire. In 1739, in debt, he sold Wimpole to the earl of Hardwicke.[74]

The maps of Wimpole Park in the Gough Collection at the Bodleian Library[75] were drawn between about 1720 and 1725 in connection with the

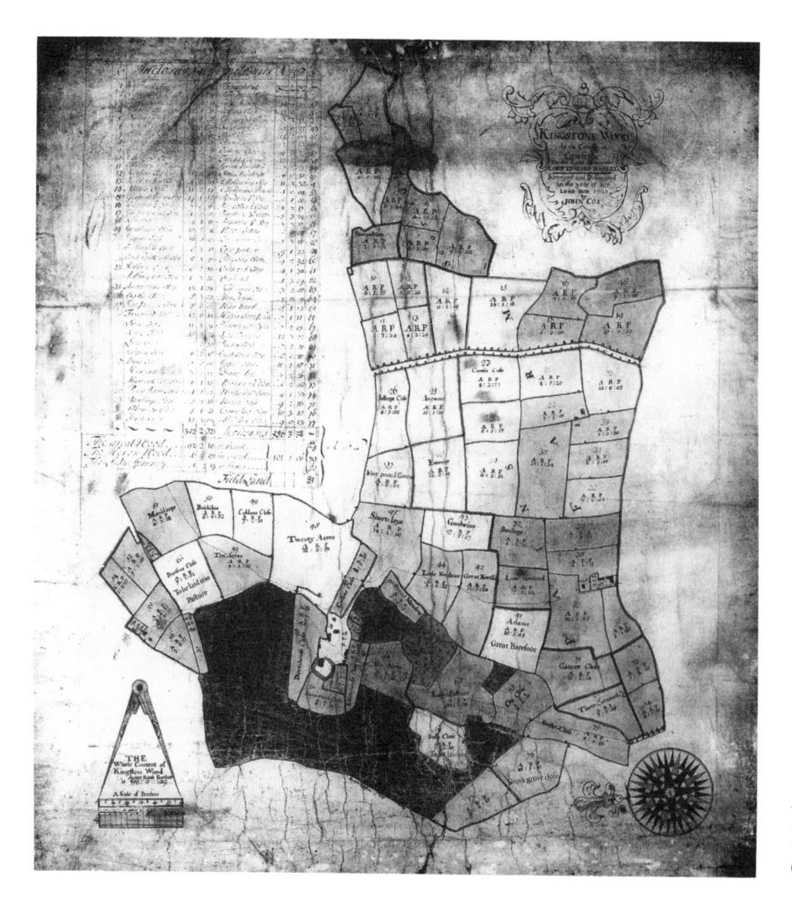

Fig. 3.15 John Cory, Kingston Wood Farm, 1720 (County Record Office, Cambridge, R52/12/5/1)

landscaping of the park by Charles Bridgeman.[76] The maps have survey marks and lines of sight and show features in the new park: avenues with the numbers of trees they were to contain, an octagonal basin, woods, fishponds, gardens, and bridges. They are drawn on paper and are undecorated and more or less uncolored. The scale is given, but no orientation is indicated and there is no formal title. The plans are not signed; they are working drawings, perhaps drawn by Bridgeman himself, a landscape gardener with a number of influential friends and patrons, and master gardener to Kings George I and George II.[77] Here are further examples of unpretentious estate maps drawn for a practical purpose, for an owner known to have interests in landscape gardening.

The 1720 map of Kingston illustrated in figure 3.15 is a contrast. With a title, clearly signed and dated, it was drawn soon after Harley acquired the estate in 1717.[78] The plan is drawn on parchment at about 5 chains to the inch and shows the farm, outbuildings, kiln, and dovecote in perspective view, and woods, the state of cultivation and fields "To be laid into Pasture," gates, orchards, a moat, hedges with hedgerow trees, and the parish boundary. The title is in a cartouche, the scale is surmounted by yellow and gray dividers, and there is a compass rose that shows thirty-two red, green, yellow, and blue points. A table gives field names, descriptions, and acreages. The map is colored: woods and pasture are green, houses are red, barns and other farm buildings are yellow, and land is outlined in green, red, blue, or brown and yellow. This is the only map known to exist by the surveyor, John Cory; it is not very sophisticated, but he was clearly not a complete novice.[79] The estate met the same fate as the land at Wimpole and was sold to Hardwicke in 1739.[80]

This last estate map is also drawn to show an estate shortly after its acquisition, commissioned partly for a practical purpose to indicate land that was to be converted to pasture and again partly for display, this time by a member of the uppermost classes of society. Harley did not employ a distinguished surveyor, and the map cannot rank among the finest and most sophisticated of its time, but it is perfectly adequate as a run-of-the-mill map of a small part of the property of a man with a reputation for no interest in the administration of his estates.

Conclusion

It has been demonstrated here that a general survey of the characteristics of estate maps, their surveyors, and the men who commissioned them, combined with a detailed examination of a few maps and their associated documents, shows that many factors influenced the shape and form of the final product. Maps were drawn to serve more than one end and could be reused over a number of years for purposes not thought of when they were originally made. To attempt to reconstruct the reasons for drawing maps, the roles they played among both private and institutional landowners, and the importance of mapmaking to the growing numbers of land surveyors who were practicing in Britain between the sixteenth and nineteenth centuries, it is essential to consider both the general and the specific historical contexts within which maps were created and used.

Notes

I am very grateful for financial support from the Economic and Social Research Council, and Emmanuel College and the Department of Geography, University of Cambridge. I should also like to thank Dr. Alan Baker for his constructive advice on this work.

The following conventions have been used in the transcription of texts: spelling and punctuation have been preserved, words in small capitals have been transcribed in lower case, italics indicates expansion of abbreviations (except for some standard ones), [] shows editorial insertions, <> shows material deleted in the manuscript, \/ shows material inserted in the manuscript, and () are parentheses in the manuscript. Dimensions of maps are measured within their borders.

In these notes the following abbreviations are used: CUL (Cambridge University Library, Map Room), ECC (Emmanuel College, Cambridge), GLRO (Greater London Record Office), PRO (Public Record Office), QCC (Queens' College, Cambridge).

1. Sarah Bendall, ed., *Peter Eden's Dictionary of Land Surveyors of Great Britain and Ireland, c. 1540–1850*, 2d ed. (London, forthcoming).

2. David Dymond, *Israel Amyce's Map of Melford Manor, 1580* (Long Melford, 1987).

3. John Harris, *The Artist and the Country House: A History of Country House and Garden View Painting in Britain, 1540–1870* (London, 1979), 47, 71.

4. Stephen Daniels, "Goodly Prospects: English Estate Portraiture, 1670–1730," in *Mapping the Landscape: Essays on Art and Cartography*, ed. Nicholas Alfrey and Stephen Daniels (Nottingham, 1990), 9–12.

5. Wiltshire County Record Office, Accession 1332, Duke of Somerset Papers, box 51. For detailed descriptions of this and all other estate maps of Cambridgeshire that will be mentioned, see Sarah Bendall, *Maps, Land, and Society: A History, with a Carto-Bibliography of Cambridgeshire Estate Maps, c. 1600–1836* (Cambridge, 1992).

6. J. P. Hore, *Sporting and Rural Records of the Cheveley Estate* (London, 1899).

7. *Dictionary of National Biography* 36:47.

8. Bendall, *Peter Eden's Dictionary of Land Surveyors*.

9. For a detailed description of the study and its findings, see Bendall, *Maps, Land, and Society*.

10. Robin Butlin, "Drainage and Land Use in the Fenlands and Fen-Edge of Northeast Cambridgeshire in the Seventeenth and Eighteenth Centuries," in *Water, Engineering, and Landscape: Water Control and Landscape Transformation in the Modern Period*, ed. Denis Cosgrove and Geoff Petts (London, 1990), 57, 65.

11. Robin Butlin, "Small-Scale Urban and Industrial Development in North-East Cambridgeshire in the Nineteenth and Early Twentieth Centuries," in *The Transformation of Rural Society, Economy, and Landscape: Papers from the 1987 Meeting of the Permanent European Conference for the Study of the Rural Landscape*, ed. Ulf Sporrong (Stockholm, 1990), 217–26.

12. Bendall, *Maps, Land, and Society*.

13. M. R. Postgate, "Field Systems of Cambridgeshire," Ph.D. thesis, University of Cambridge, 1964.

14. Philip Jenkins, "Cambridgeshire and the Gentry: The Origins of a Myth." *Journal of Local and Regional Studies* 4(1984):1–17.

15. Edward Walford, *The County Families of the United Kingdom* (London, 1860); H. A. Clemenson, *English Country Houses and Landed Estates* (London, 1982), 21–23.

16. Postgate, "Field Systems of Cambridgeshire," 149–50.

17. Bendall, *Maps, Land, and Society*.

18. The following three sections are summaries of the discussion in ibid.

19. Nicholas Cox, *Bridging the Gap: A History of the Corporation of the Sons of the Clergy over 300 Years, 1655–1978* (Oxford, 1978), 56–57.

20. *The Victoria History of the Counties of England: Cambridgeshire* (1978), 6:193; (1990), 9:404.

21. GLRO A/CSC/1101 for map of Wisbech area, 1716; CUL MS Plans 713 for map of West Wratting, 1719; CUL MS Plans 716 for map of West Wratting and Willingham, 1719.

22. A. Stuart Mason, *Essex on the Map: The Eighteenth-*

Century Land Surveyors of Essex (Chelmsford, England, 1990), 39; Bendall, *Peter Eden's Dictionary of Land Surveyors.*

23. Cox, *Bridging the Gap,* 62.

24. Bendall, *Peter Eden's Dictionary of Land Surveyors.*

25. British Library, Maps R.a.2.

26. E. H. Pearce, *The Sons of the Clergy: Some Records of 275 Years* (London, 1928), 121.

27. CUL MS Plans 717; GLRO/CSC/1101/1–15, by W. P. Attfield of Whetstone and Hadley, Middlesex.

28. Bendall, *Maps, Land, and Society.*

29. QCC Conclusion Book 1733–87, 71r, 72v.

30. QCC Book 48, 7.

31. G. E. Aylmer, "The Economics and Finances of the Colleges and University, c. 1530–1640," in *The History of the University of Oxford,* vol. 3, *The Collegiate University,* ed. James McConica (Oxford, 1986), 534–35.

32. C. L. Shadwell, *The Universities and College Estates Acts, 1858–1880: Their History and Results* (Oxford, 1898), 11–12.

33. J. P. D. Dunbabin, "College Estates and Wealth, 1660–1815," in *The History of the University of Oxford,* vol. 5, *The Eighteenth Century,* ed. L. S. Sutherland and L. G. Mitchell (Oxford, 1986), 273–86.

34. QCC Account Book "Coll. No. 2," 3–5.

35. Bendall, *Maps, Land, and Society.*

36. QCC book, 48, 49.

37. QCC Book 8.

38. QCC Box 104, Account paid March 1806.

39. QCC Box 104, Account paid 4 January 1811.

40. QCC Book of Orders 1787–1832, 12 January 1822.

41. QCC Box 123; Books of Orders 1787–1832, 3 March 1827.

42. Bendall, *Maps, Land, and Society.*

43. Trinity College, Cambridge, Shelf 85.3.

44. *Victoria History* (1953), 4:138–39.

45. QCC Box 114, Letter for Alexander Watford to Bursar, 9 September 1839.

46. QCC Box 116, Letter for Alexander Watford to Charles Day, 18 June 1840, and note to Bursar.

47. *Cambridge Chronicle,* 21 July 1764.

48. Bendall, *Peter Eden's Dictionary of Land Surveyors.*

49. G. R. Owst, "Iconomania in Eighteenth-Century Cambridge: Notes on a Newly-Acquired Miniature of Dr Farmer and His Interest in Historical Portraiture," *Proceedings of the Cambridge Antiquarian Society* 42 (1949):80.

50. Bendall, *Maps, Land, and Society;* PRO PROB 6/175/398r; *Gentleman's Magazine* 69, pt. 1 (1799):260.

51. Bendall, *Peter Eden's Dictionary of Land Surveyors.*

52. QCC Box 104, Account to Administratrix of Alexander Watford, paid on 3 March 1802.

53. PRO PROB 6/177/436r.

54. QCC Box 104, Account paid 3 April 1802.

55. Bendall, *Peter Eden's Dictionary of Land Surveyors.*

56. Bendall, *Maps, Land, and Society;* PRO PROB 6/220/fol. 332v; *Gentleman's Magazine* 22 (1844):329.

57. ECC Box 27.C1–2.

58. ECC BUR.8.4, Payments for 8 November 1745 and 29 May 1746.

59. ECC COL.14.2 fol. 23v.

60. ECC BUR.8.5, Payment for 18 May 1754; ECC BUR.0.3a, Account paid 28 October 1777; ECC BUR.0.3a, Account paid 3 August 1778.

61. ECC FEL.5.1.

62. ECC BUR.0.4(b), Account paid 25 October 1815.

63. Bendall, *Maps, Land, and Society.*

64. *Victoria History* (1978), 6:85.

65. See maps in the archives of Clare and St. John's Colleges; Bedfordshire County Record Office, R1/153–54.

66. J. A. Venn, *Alumni Cantabrigienses: A Biographical List of All Known Students, Graduates, and Holders of Office at the University of Cambridge from the Earliest Times to 1900; Part 2: From 1752 to 1900* (Cambridge, 1940–54), 4:7.

67. GLRO ACC.1876/MP2/2.

68. *Victoria History* (1978), 6:40.

69. Ibid., 89.

70. Cambridge University Library, Department of Manuscripts, Palmer MS C.9.

71. *Victoria History* (1978), 6:48.

72. O. R. F. Davies, "The Wealth and Influence of John Holles, Due of Newcastle, 1694–1711," *Renaissance and Modern Studies* 9(1965):22–46.

73. *Victoria History* (1973), 5:265.

74. *Dictionary of National Biography* 24:394.

75. Bodleian Library, Oxford, MS.Gough Drawings.a.4, fols. 30, 31, 35, 69.

76. Royal Commission on Historical Monuments, England, *An Inventory of Historical Monuments in the County of Cambridge,* vol. 1, *West Cambridgeshire* (London, 1968), 215–16, plate 121; David Souden, *Wimpole Hall, Cambridgeshire* (London, 1991), 14–15, 80.

77. *Dictionary of National Biography, supplement,* 3:268–69; Bendall, *Peter Eden's Dictionary of Land Surveyors.*

78. *Victoria History* (1973), 5:114.

79. Bendall, *Peter Eden's Dictionary of Land Surveyors.*

80. *Victoria History* (1973), 5:114.

Chapter FOUR

THE ESTATE MAP *in the* NEW WORLD

DAVID BUISSERET

Introduction

Military and administrative practices, developed to control subject populations in the Old World, were often carried across the Atlantic. Thus the early Spanish and Portuguese colonizing methods adopted for the Atlantic islands—the Azores, the Madeiras, and the Canaries—found a wider application in Central America and Brazil, and the English experience in Ireland offered lessons for the eventual settlements on the east coast of North America.[1] In Ireland the English invaders had made extensive use of estate maps, as we have seen, and this mode of control soon spread to the islands that they seized in the Caribbean: St. Christopher, Barbados, Antigua, Jamaica, St. Vincent, and so forth. The seventeenth-century French did not have a metropolitan tradition of estate maps (see chapter 1), but they too soon began using them for their Caribbean possessions, perhaps in imitation of their English contemporaries. Estate maps did not emerge in the Spanish, Dutch, or Danish islands, for reasons that will become clear as we examine their social and economic structure.

On the mainland the estate map in its traditional form became well established in South Carolina during the eighteenth century, no doubt in part because of that region's close ties with islands like Jamaica.[2] Planters in many other states of the South also used estate maps, though they were nowhere as widespread as in South Carolina. There were other types of topographical map in the anglophone regions; the Shakers drew them in Pennsylvania, for instance, and so did the owners of the patroonships on the Hudson River. As we shall see, though, these maps differed in crucial respects from the typical estate maps of the South.

Outside the anglophone regions, too, large-scale rural maps came into existence. Some of the seigneuries of the St. Lawrence River were mapped in the seventeenth century, as were many of the ranches of California in the eighteenth and nineteenth centuries. In the case of these California ranch maps, or *diseños*, the prototypes are probably to be found in Mexico, where there was a tradition of large-scale mapping going back to the Spanish conquest and indeed to pre-Columbian times. One way and another, virtually all the areas of Central and North America occupied by the Europeans before 1800 had developed some way of delineating their countryside at a relatively large scale.

The "British" Islands of the Caribbean

Parties from England reached St. Christopher and Barbados in the 1620s, and by the 1650s it was beginning to be clear that sugar would be a profitable crop. The land therefore began to be valuable, and the earliest estate plats of Barbados date from this time.[3] A number survive in the island's archives,[4] but they are by no means abundant, perhaps because the planters of Barbados tended to direct relatively minor enterprises, and to live on the island.

Among the other islands, Jamaica represented the opposite extreme of economic and social conditions, with vast estates and often absentee owners; the island produced by far the largest number of estate maps, described in chapters 5 and 6. Scattered material survives from many of the smaller islands as well, however, and figure 4.1 shows us an example. The original plan was compiled in 1811 for James Wilson by Joseph Billinghurst, and shows Cane Grove Estate on the southwest coast of the island of St. Vincent, about 4 miles north of the capital, Kingstown. The original, which is in the Boston Public Library, is too large (roughly 24 inches by 36 inches) to be reproduced.

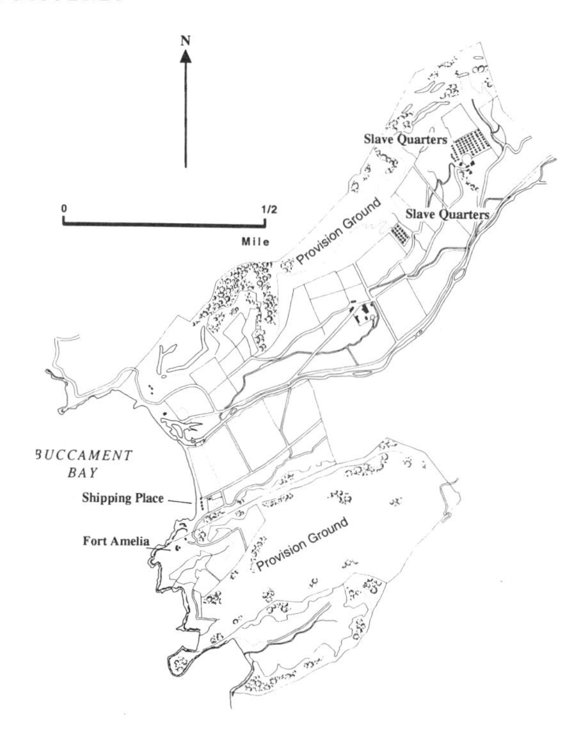

Fig. 4.1 Redrawing of a map by Joseph Billinghurst, Cane Grove Estate (St. Vincent), 1811 (Boston Public Library)

Billinghurst is able to show all the cane fields and their size (though he disappointingly does not tell us their names). He notes the slaves' provision grounds, in the marginal land, and shows us in a detail where they and the master lived: cheek by jowl, as was the custom. The other detail sets out the plan of the works where the sugar was processed, giving the location of the water mill, boiling house and curing house. The roads leading down to the "publick shipping place" are also shown; in short, we have all the details needed to reconstruct the spatial arrangements of the estate, for this map compares very closely with the Directorate of Overseas Surveys map of the same area.[5] Buccament Bay also shipped sugar from Pembroke Estate, whose "shipping road" is shown; this loading area was protected after a fashion by the guns of Fort Amelia, one of the omnipresent Caribbean fortifications designed to drive off privateers. In summary, Billinghurst's map is typical of estate maps produced during the eighteenth and nineteenth centuries, which on the whole have not been fully used by historians as they deserve.

1 Adrien de Montigny, view of Halluin, c. 1600, from *Les Albums de Croy*, ed. Jean-Marie Duvosquel, vol. 12 (Brussels, 1985), plate 63. This view, framed by elegant floral drawings in a well-established late medieval style, would have reminded the duc de Croy of the general appearance of Halluin; its purpose was more celebratory than functional.

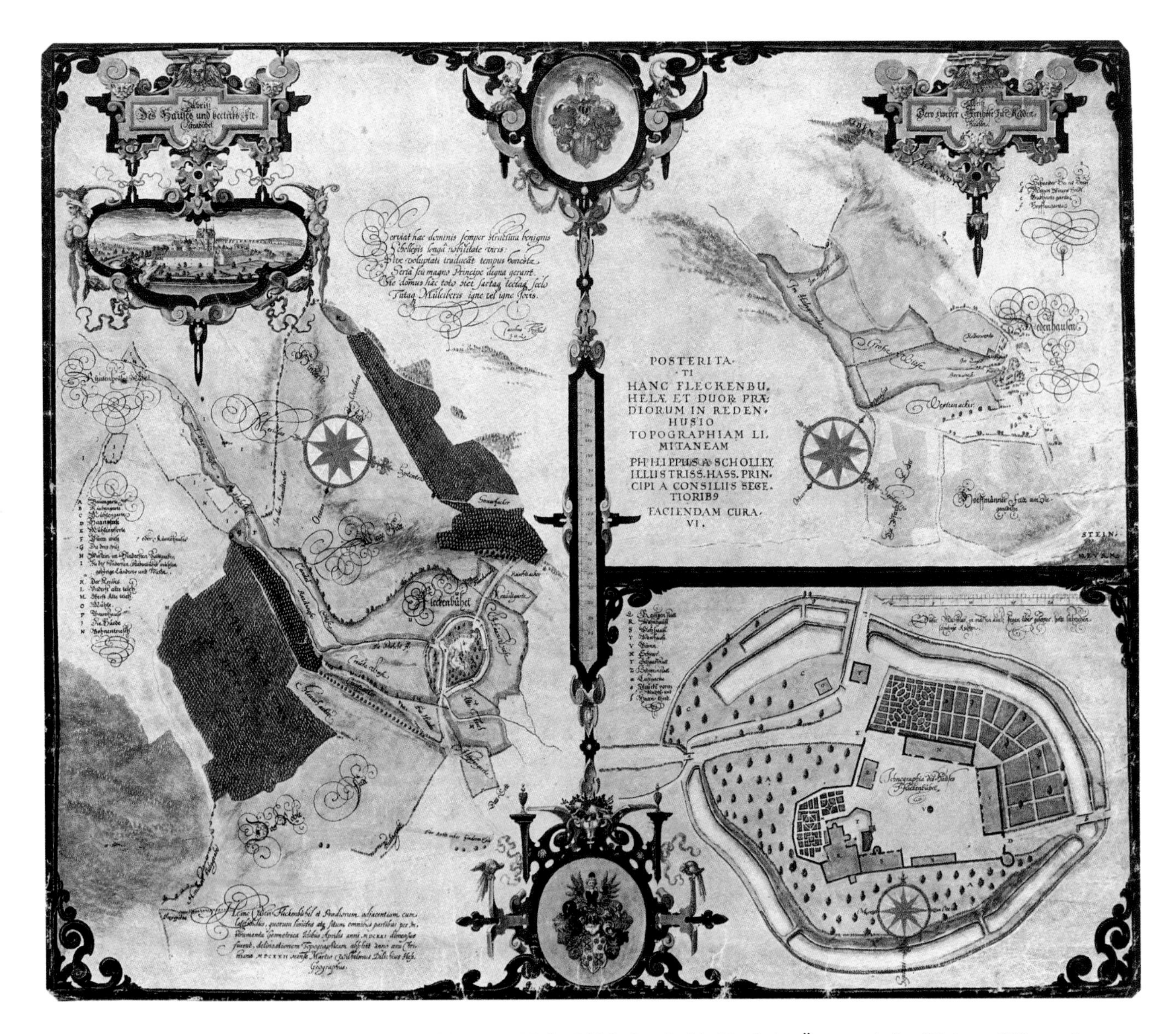

2 Wilhelm Dilich, *Schloss und Bezirk Kleckenbühl*, from *Wilhelm Dilichs Landtafeln Hessischer Ämter zwischen Rhein und Weser*, ed. Edmund Stengel (Marburg, 1972). This very unusual set of images of the castle and estate at Kleckenbühl shows how skilled some cartographers in Germany had become by the early seventeenth century; Dilich uses a variety of orientations and scales to bring out the salient features.

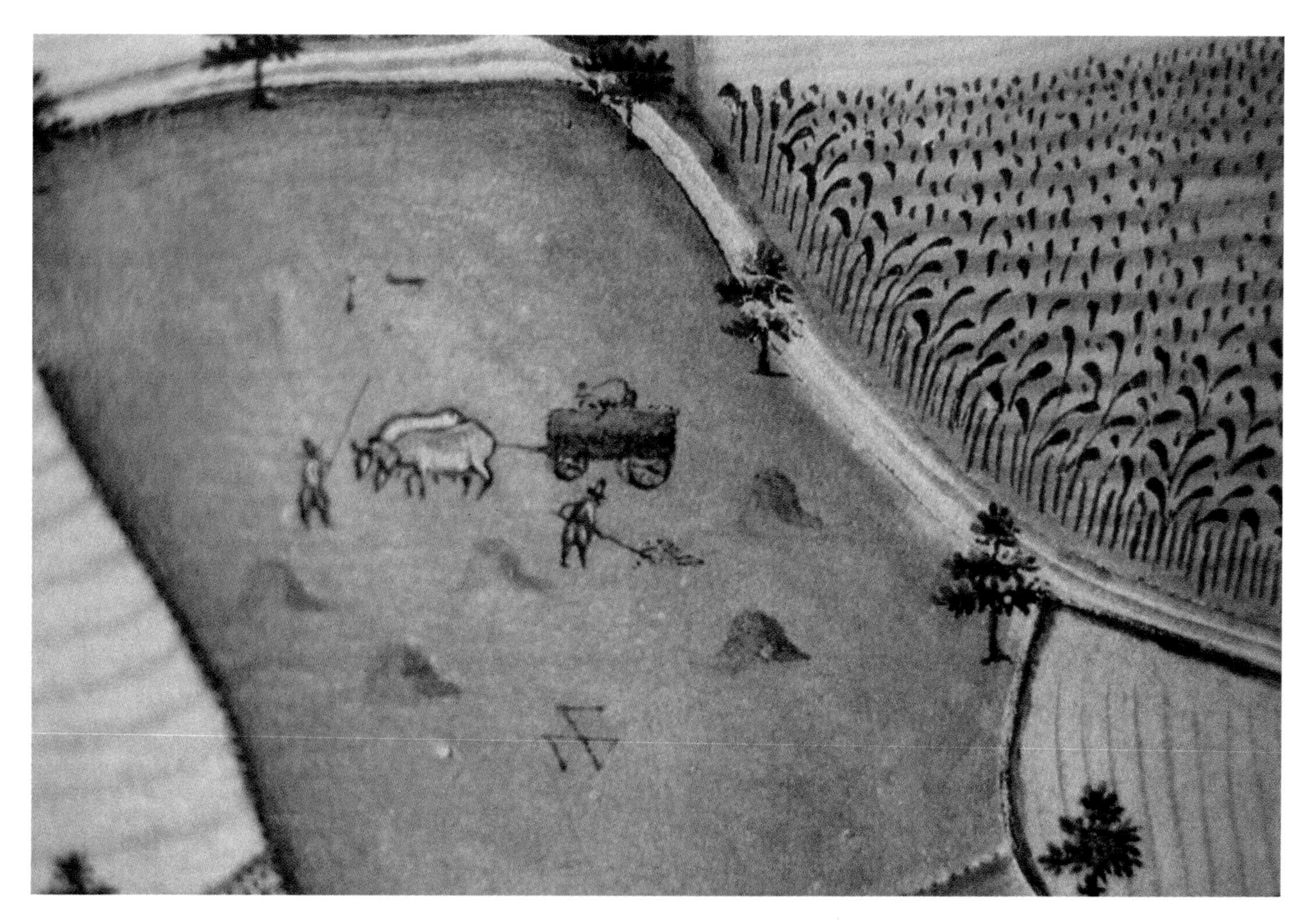

3 Mark Pierse, detail from "Plot and Description of the Lordship of Pirton," 1623 (by permission of the trustees of the Croome Estate, Worcester). Scenes like this allow the student to peel back the layers of change in the English countryside: oxen, not horses, pull the cart, and the road is gravel rather than tarmac. Otherwise, the basic techniques shown here were in use until the twentieth century.

4 Anonymous, plat of land belonging to Governor Joseph West, 1680 (South Carolina Historical Society). This lovely map comes from a time when the celebratory aspect of estate maps was still strong in the New World; the later estate maps in South Carolina would be less elegant, though no less accurate, and more informative.

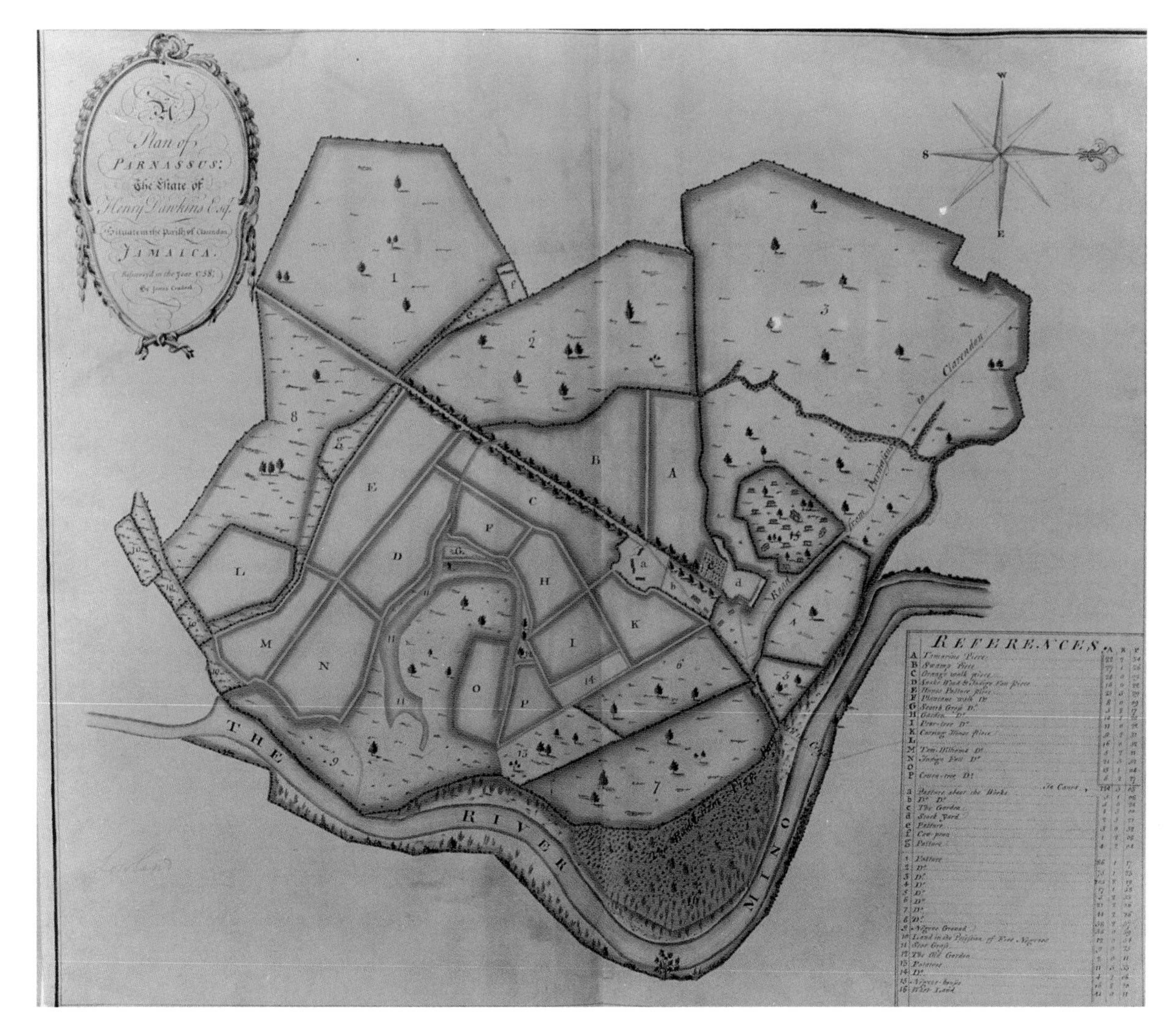

5 James Cradock, plan of Parnassus Estate, 1758 (National Library of Jamaica). This beautifully finished plan comes from the Dawkins Plantation papers, which contain other, similar plans. Images like these may be compared very tellingly with the existing features of the Jamaican countryside.

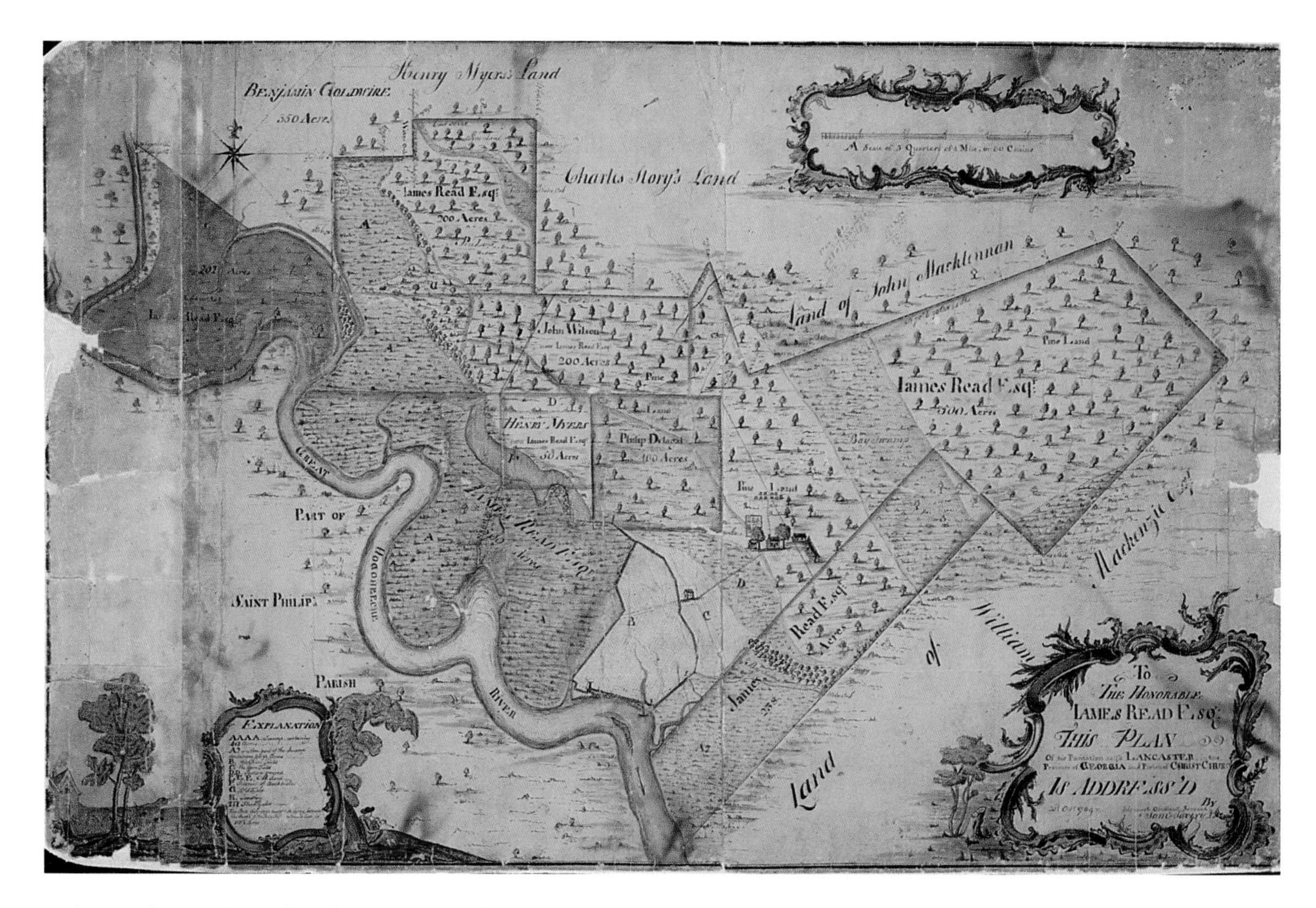

6 Samuel Savery, James Read's estate on the Ogeechee River, 1769 (U.S. National Archives). This map should be compared with the USGS map shown on page 107; one is struck not only by the accuracy of Savery's delineation, but also by the way in which an apparently promising tract of land has failed to live up to the hopes of its owner.

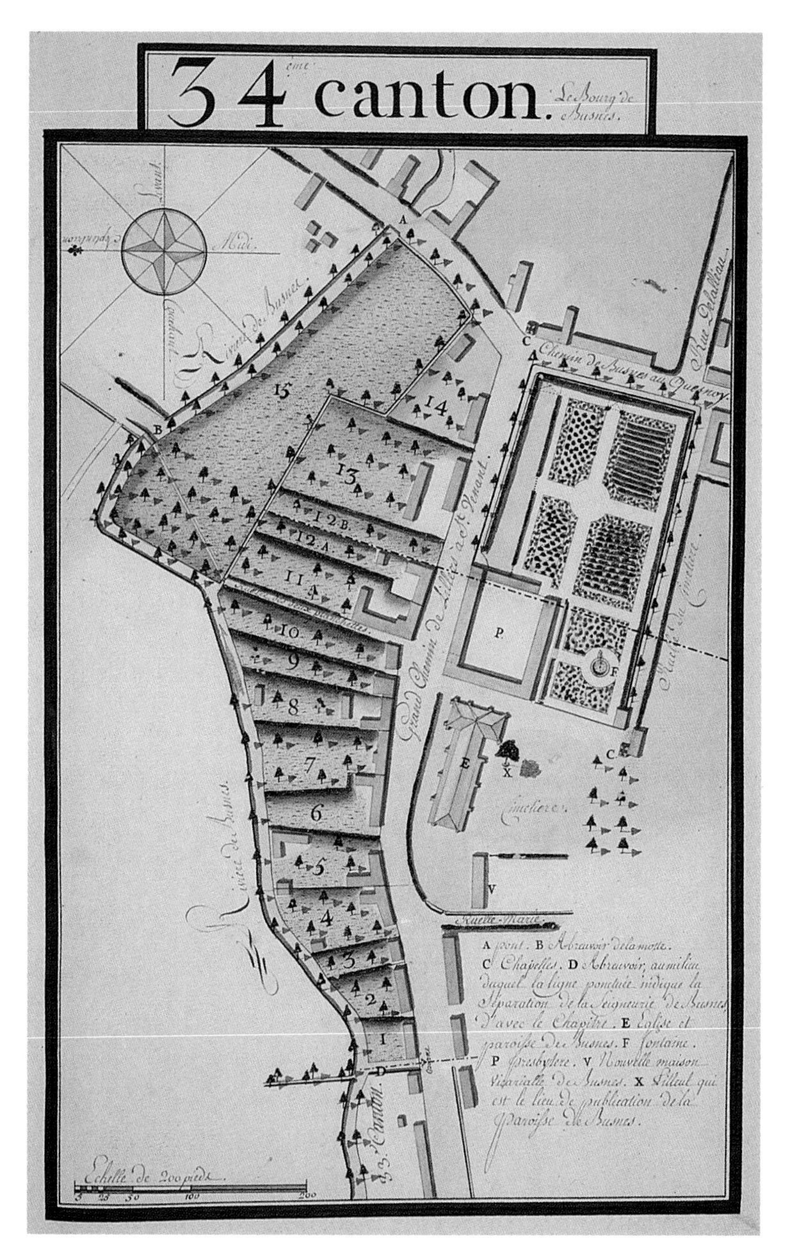

7 Christophe Verlet, map from the atlas of the parish of Busnes, near Lille (France), 1782 (Newberry Library). Christophe Verlet's atlas contains almost a hundred other maps like this one, showing the parish in great detail. Such atlases were important tools for "improving" landlords in late-eighteenth-century France, and had counterparts elsewhere in Western Europe.

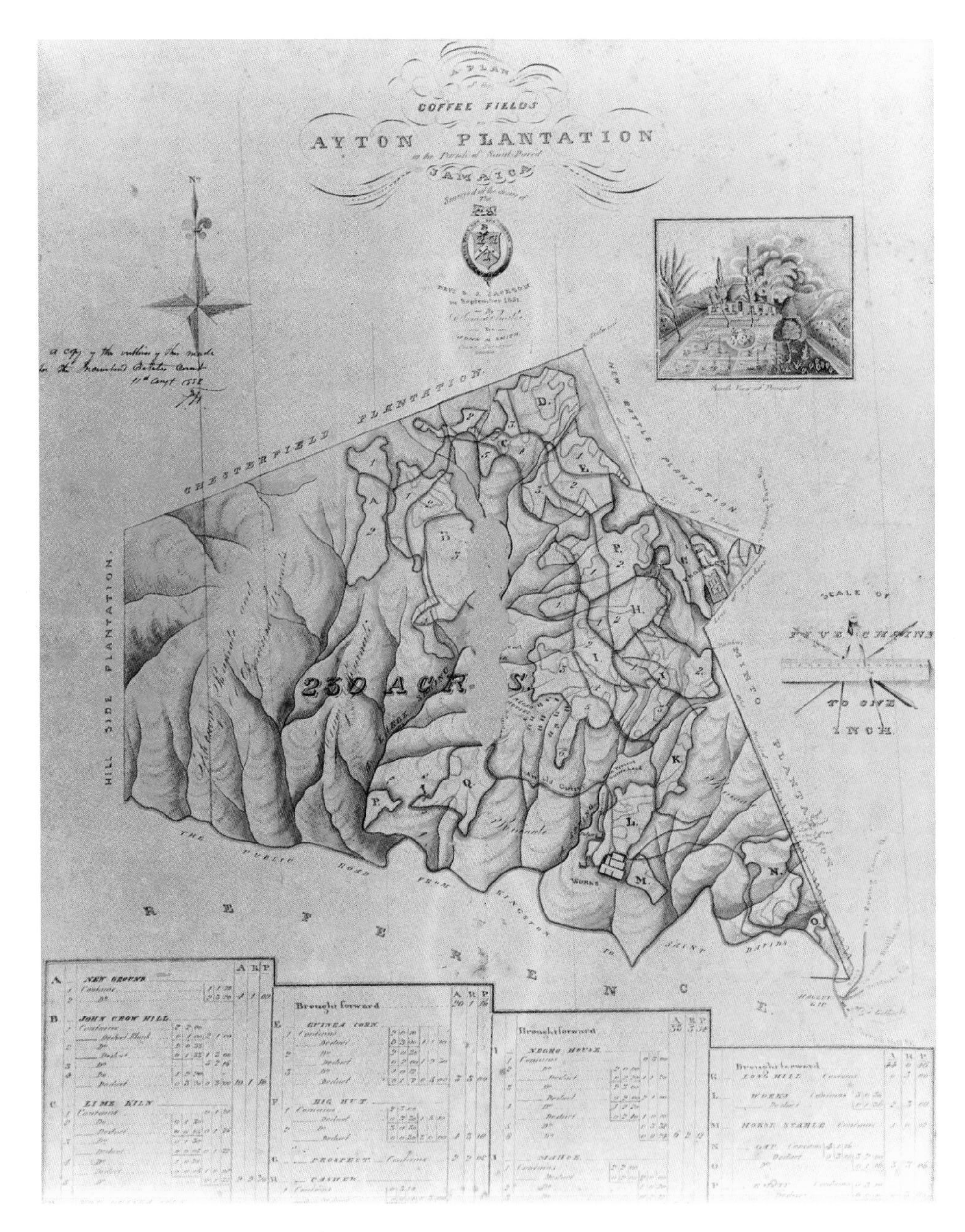

8 James Smith, plan of the coffee fields on Ayton Plantation, 1851 (National Library of Jamaica). We have here an example of mapping precipitous country in the mountains of Jamaica. Such maps were very carefully drawn, in spite of the difficulties of the terrain, in order to show the small and irregular plots upon which the valuable coffee crops would be grown.

The French Possessions

In the French territories, estate maps became relatively abundant during the eighteenth century,[6] ranging very widely in quality of execution. One of the crudest and most evocative of these plans is the one that Etienne-François Turgot had drawn for his estate in Cayenne (figure 4.2). Turgot (1721–89) was a brother of the well-known financier Anne-Robert-Jacques Turgot, and about 1770 was appointed governor of French Guiana, where the French crown hoped to establish a new colony. The accounts for this venture survive in the Bibliothèque Nationale at Paris[7] and tell a tale of disaster, with the climate resisting all attempts at cultivation and even at survival. Our map speaks eloquently of the problems of this venture; instead of well-regulated fields and buildings, we have crudely delineated circular patches of the red dye *annotto* ("roucou"), surrounded by bush, giving onto a creek. The eleven slave huts form a bedraggled line, and there seems to be an attempt at a formal garden, down by the water. Looking at this little map, we are not surprised to learn that the whole venture was a dismal failure.

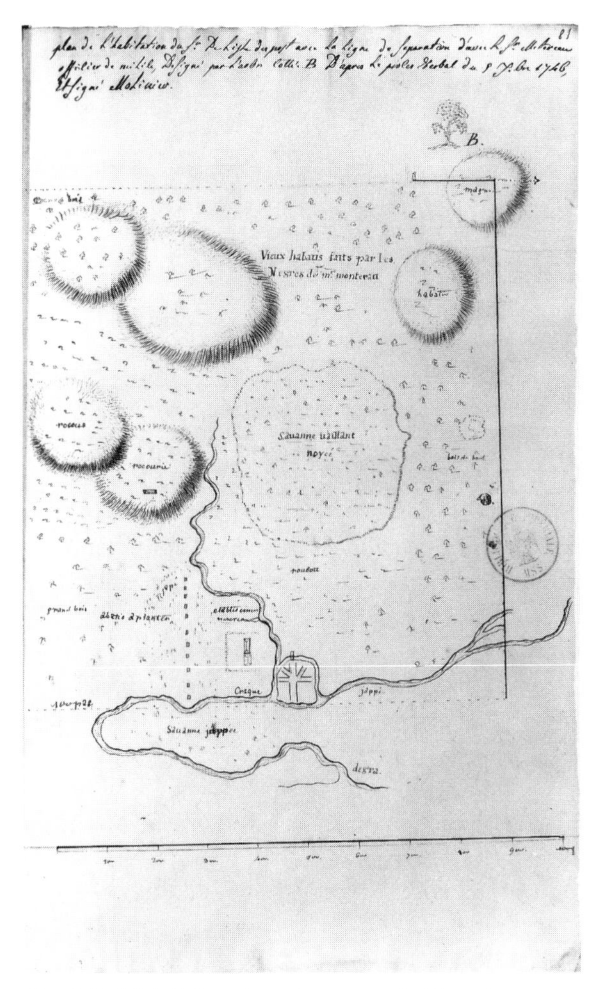

Fig. 4.2 N. Molinier, plan of an estate in French Guiana, c. 1770 (Bibliothèque Nationale, Paris)

Much more prosperous plantations were of course established on Martinique and Guadeloupe during the seventeenth century, while the estates of St. Domingue were by the end of the eighteenth century the most productive in the whole Caribbean. Figure 4.3 shows the late-eighteenth-century plan of a relatively modest estate belonging to "M. de Lacaze," in the Artibonite region of central western St. Domingue.[8] Oriented westward, it has a scale and a key to no fewer than fifty-eight numbered features. Forty numbers refer to the fields, generally of uniform size and bearing optimistic names such as L'Intarissable (the ever-abundant), La Fertile, La Superbe, and so forth. In a large enclosure by the fields are the estate buildings: the slave quarters (with a modest sixteen houses), the great house, the hospital, the forge, and so on. A stream winds its way across the map, and the names of adjacent proprietors are noted.

Altogether more impressive, both in execution and in the size of the area shown, is the "plan de la Première, Seconde et Troisième Habitations de M. de Laborde" (figure 4.4). This huge establishment

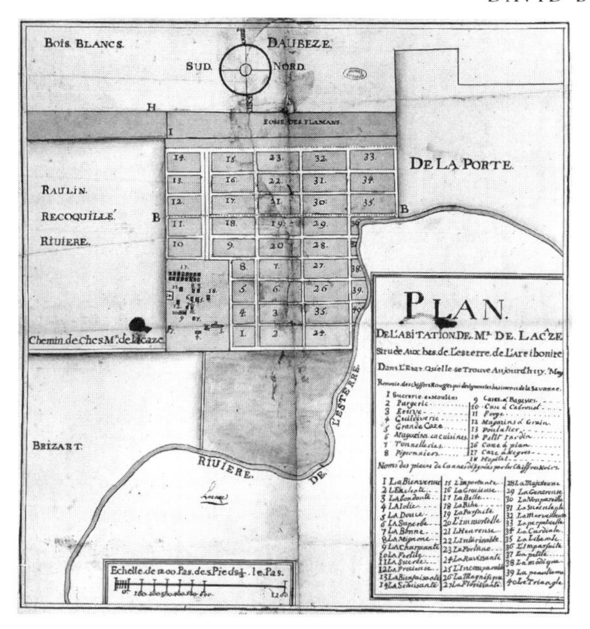

Fig. 4.3 Anonymous, *Plan de l'habitation de M. de Lacaze*, c. 1770 (Archives Nationales, Paris, St. Domingue, N III 17)

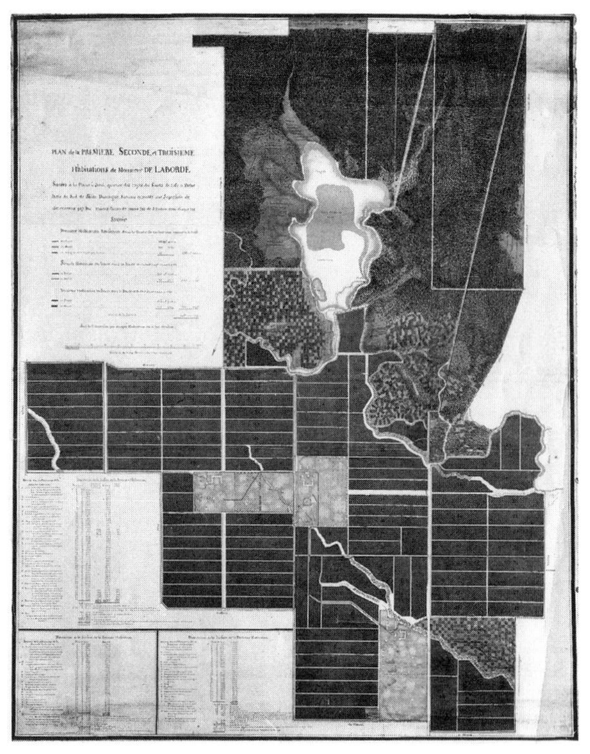

Fig. 4.4 Anonymous, plan of the plantations of M. de Laborde, c. 1790 (Archives Nationales, Paris)

lay on the Plaine à Jacob, to the north of Les Cayes on the southern coast. Laborde was a banker, whose three contiguous plantations, worked by well over one thousand slaves, were among the largest on the island in 1791.[9] We do not know who drew the plan, but it is remarkably professional in all its details and covers a huge area, roughly 3 miles broad by 5 miles high. Indeed, the map is so hard to reproduce that I have provided a drawing of it (figure 4.5). Here we can see the buildings for each of the three plantations, each containing slave quarters ranged around the side of a rectangular area (they seem formerly to have been aligned in rows in the middle of this area). Within this area are the great house, the mill, the hospital, and so forth. By a lake up in the mountains the mills are driven and the fields watered; up there, too, are the slaves' provision grounds. Tracks run between the fields, and major roads connect the whole vast enterprise with the outside world. The evidence of this magnificent plan makes it easy to understand not only why St. Domingue had become the leading producer of sugar, but also why the eventual social revolution was on such a massive scale.

It is, unfortunately, impossible to offer any serious estimate of the number of such plans produced for St. Domingue. No recent work has been carried out on this subject, and such a venture would call for wide-ranging investigations, not only in the French archives but also in private collections on both sides of the Atlantic. Nor is it possible to assess the number of surveyors at work, or the nature of their training. It is very unlikely that there were

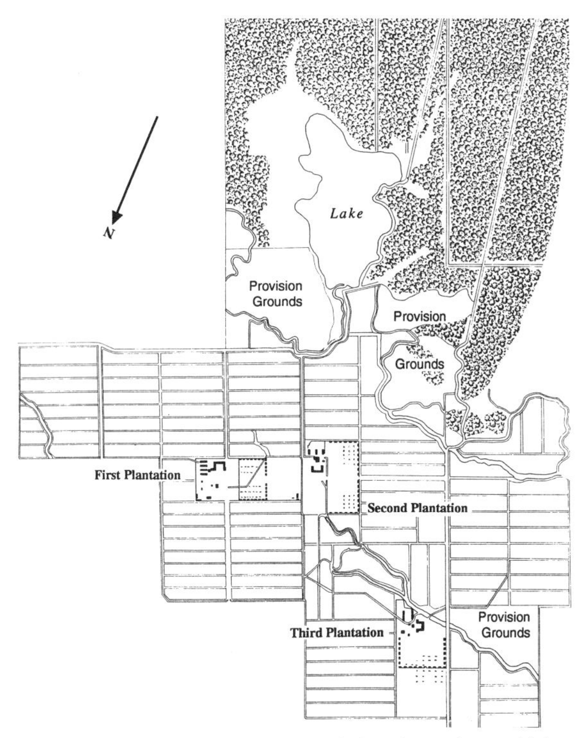

Fig. 4.5 Redrawing of the plan of the plantations of M. de Laborde

as many as on the island of Jamaica, where they have left ample evidence. Perhaps on the French islands surveys were often carried out by military engineers; certainly the map reproduced by Gabriel Debien in *La Sucrerie Galbaud du Fort* was the work of a "dessinateur des fortifications de Saint-Domingue."

The Spanish Islands

Of all the European powers, the Spaniards arrived first in the Caribbean and longest controlled the major islands; it might thus have been expected that they would have early developed detailed topographical maps of them. It did not work out that way, however, even though during the later sixteenth century the Spanish crown did commission detailed maps of many of its overseas possessions.[10] This absence of mapping activity in the islands was partly the result of Spain's increasing neglect of them, after more exciting possibilities opened up on the mainland after 1519. However, it was also the result of a distinctive system of land distribution in the Spanish Greater Antilles.[11]

In Cuba, Santo Domingo, and Puerto Rico, land outside the cities was granted to individuals or corporations by virtue of royal *mercedes*, generally assigned by the municipalities. These *mercedes* did not refer to precisely surveyed areas, but rather to localities. Thus in Santo Domingo grants were made in terms of the number of *montones*, or plantings, that came under the control of subordinate *caciques*. In Cuba the system seems to have been somewhat more spatially directed; thus cattle ranches, or *hatos*, were assigned a circle of land within a radius of 2 leagues (roughly 6 miles) from a central point, and hog ranches, or *corrales*, a circle with a radius of 3 miles.

In time this method of allocation in Cuba led to grave problems, since some of the circles intersected, and in others hog ranches fell wholly within cattle ranches. In the other islands, too, land grants based on areas occupied by the indigenous inhabitants proved problematic, as the native population dropped in a catastrophic way. One way and another, these Spanish systems did not lead to any need for a detailed system of topographical mapping, particularly as the general population of the islands tended to fall precipitately; for instance, whereas there were 53,000 Spaniards on Santo Domingo in 1519, by 1730 there were only 6,000. It would not be before the nineteenth century that detailed topographical surveys would be carried out in the large Spanish islands.

The Dutch and Danish Islands

The Dutch began invading the Spanish preserves in the Caribbean during the 1620s, and eventually established themselves on a number of islands. Most were small, only Curaçao being an island of any size. Here the Dutch developed a thriving city at Willemstad, and eventually filled the surrounding countryside with elegant houses. These houses were mere merchants' retreats, however; they had very little in the way of agricultural activity associated with them, and consequently generated nothing like estate plans.[12]

The Danish West India Company bought St. Croix from the French in 1733, and soon began mapping this small island.[13] The manuscript map produced by Johan Cronenburg and Johan van Jaegersburg during the 1740s was very detailed, and gives a full picture of the progress of Danish settlement, down to the location of slave quarters, plantation houses, and mills of different types. This map was superseded in 1754 (the year that the island reverted to the Danish crown) by the printed map of Jens Michelsen Beck. Although less full and accurate, this became the standard reference map for the island, whose individual landowners do not seem to have produced maps. Perhaps the island was too small for that, or perhaps the Danish planters were happy simply to use the work of government surveyor Beck, as their counterparts back in Denmark used official surveys.

The East Coast of North America: The States between New England and North Carolina

From quite early in the process of their settlement, the English in New England commissioned plats of their land, and some of them were quite elaborate. For instance, in 1701 William Goodsoe drew *A Platt of Mr. Humphrey Chadburn's Farm att Sturgen Creek;* it has a magnificent wind rose and a convincing delineation of the house, and sets out the bounds of the property alongside Sturgeon Creek.[14] The surveyors never progressed from these plats into the compilation of estate maps, however; as Peter Benes puts it in *New England Prospect,* "the English-style 'estate map,' a property plan whose purpose was both decorative and administrative, did not become an accepted mapping form in New England, where small holdings were the rule and tenancies on large properties the exception."[15] Even the great patroonships of the Hudson River valley do not seem to have generated estate maps, which are absent both from the New York State Archives and from the holdings of the Maryland Historical Society.[16]

The situation in Virginia is more complicated. This state had many substantial landowners, particularly along the James River. Surveyors like John Warner were active there in the early eighteenth century, working for patrons familiar with English and West Indian models of estate maps.[17] Yet they seem to have drawn very few of them,[18] and those that they did draw were very low-key, not at all suitable for display.[19] The Virginia surveyors tended to be amateurs, who often became leading members of society; they needed rapid but not necessarily very sophisticated results.[20]

Sarah Hughes puts the problem this way:

> No evidence exists of such [estate maps] made in the first two periods of surveying in Virginia, though they may once have existed. Virginians at an early date were exceptionally conscious of the insignia of rank and property . . . So it is likely some would have been willing to pay a surveyor to produce a prestigious drawing of their plantation to decorate their

> wall or show to envious relatives at home in England. Yet, if this did happen, it is curious that the first extant Virginia plat is not especially decorative.[21]

The most extensive—and indeed unique—set of surviving estate maps is that drawn by George Washington of his estate on the Potomac River between 1766 and 1793. In 1766 he drew *A Plan of My Farm on Little Hunting Creek and Potomack River* (figure 4.6).[22] It is a very workmanlike map, showing the central farm buildings, the access roads, and the fields, some of which contain orchards. Washington continued all his life to draw maps and in 1793 composed what he called "a rude sketch of the farms." Figure 4.7 shows this map, whose original is now at the Huntington Library in California. Far from being a rude sketch, it is in fact a most accomplished delineation of the five farms that by then made up his estate. Washington added notes of different types; Dogue Runn Farm, for instance, had "land capable of high improvement into meadow." At *B*, he notes, "is a most beautiful site for a gentleman's seat." His own seat, of course, was Mount Vernon, on Mansion House Farm; many of the notes suggest that he had in mind to sell other parts of the land for house sites. As well as being an interesting commentary on the growth of Washing-

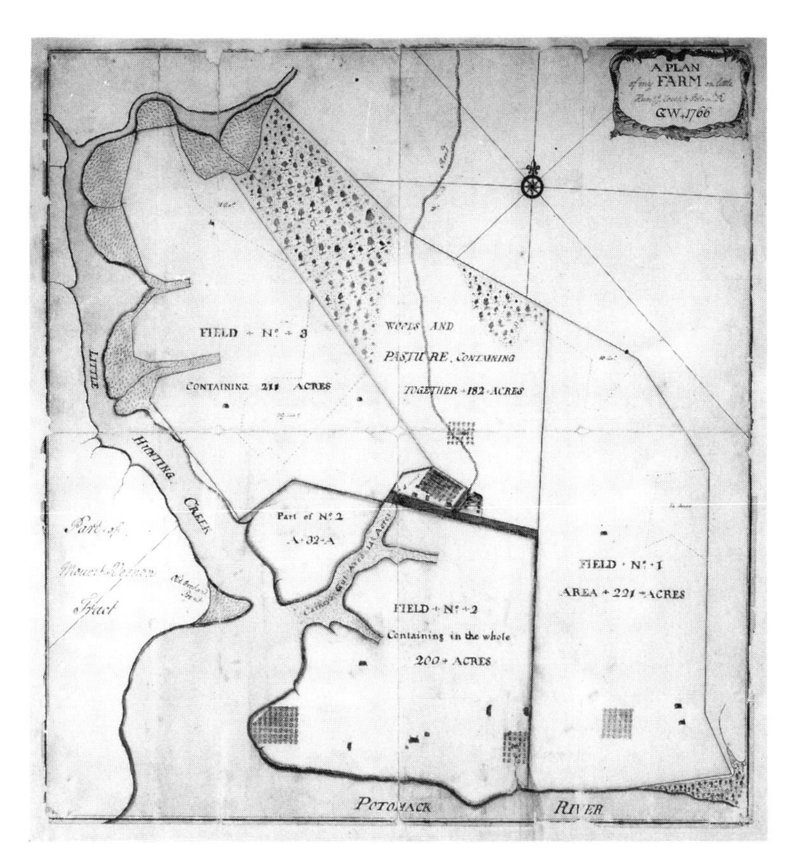

Fig. 4.6 George Washington, *A Plan of My Farm on Little Hunting Creek and Potomac River,* 1766 (Library of Congress)

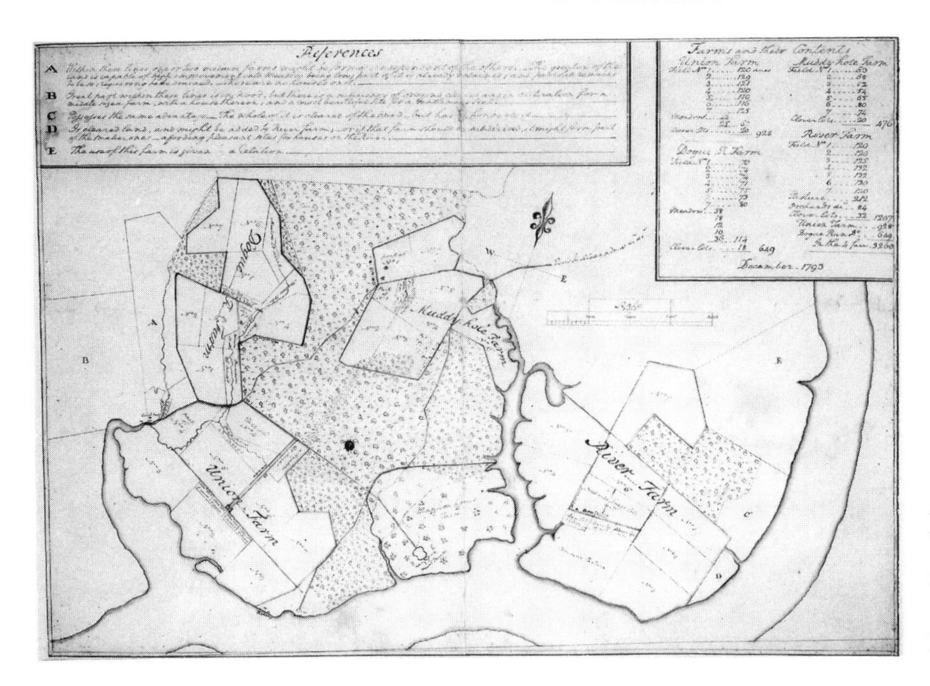

Fig. 4.7 George Washington, plan of the farms at Mount Vernon, 1793 (Huntington Library, HM 5995)

ton's lands from the original River Farm in figure 4.6, this map is also a fine example of a landowner taking stock of his holdings in cartographic form, and reflecting on how best to exploit them.

No doubt this mapping skill was relatively widespread among the landowners of the early republic; Thomas Jefferson too was an accomplished cartographer, though his work did not include estate maps.[23] In the end, Washington's five farms were swallowed up by suburbia. Figure 4. 8 shows how they look today. Mansion House Farm contains a huge traffic circle on Memorial Highway, and the very outline of most of the other farms seems to have been lost. Only in the case of Union Farm have some of the old tracks survived, as modern roads. It seems particularly unfortunate that the developers did not think of using the names of any of Washington's farms for their subdivisions, which might have served as a reminder of his former presence here.

South Carolina

In South Carolina the most propitious conditions existed for the emergence of estate maps, for here the landowners of the great riverine estates were familiar with West Indian practices, and could call upon the services of numerous surveyors from the time of Florence Sullivan (surveyor general 1670–71) onward.[24] The earliest plans show indigo estates, but from about 1720 onward rice was the main crop, until by 1839 three-quarters of the country's production came from South Carolina.[25] After 1860 rice production began to give way to cotton, and the hurricanes around 1900 dealt a final blow to most of the rice fields.

This rice economy is delineated in a great abundance of estate plans, preserved in Charleston at the Historical Society of South Carolina and at the Office of Records and Mesne Conveyances (McReady Plat Collection), and in Columbia in the South Caroliniana Collection of the University of

South Carolina.[26] About 270 surveyors were at work between 1670 and 1775, to judge from Pett-Conklin's figures;[27] she distinguishes between the aristocratic, dynastic surveyors of Virginia, and the socially less prominent ones of South Carolina. A crude and preliminary survey of the Charleston material turned up fourteen surveyors whose work survives in at least twelve examples.[28] Some of them, like Ephraim Mitchell, Joseph Purcell, Henry Ravenel, and John Wilson, were also authors of printed, small-scale maps. These surveyors—one could almost say this school of surveyors—would merit close examination. Some of them seem to have French connections; was Peter Belin, for instance, related to the Bellin family of cartographers? Others, pace Pett-Conklin, seem to belong to surveying dynasties, like Charles and Isaac Gaillard, who surely must be related to the Tacitus Gaillard who in 1770 produced a manuscript map of South Carolina.

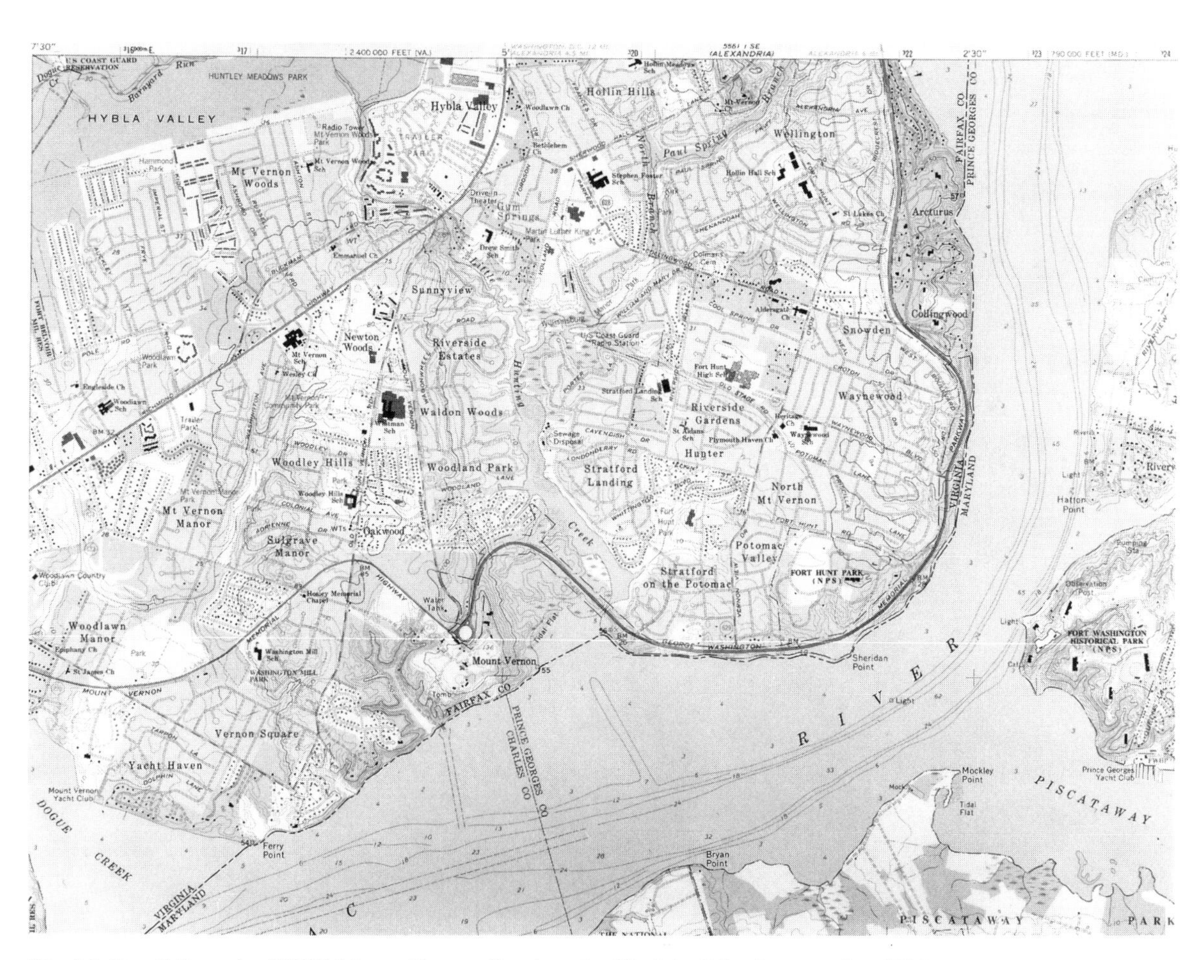

Fig. 4.8 Detail from the USGS Mount Vernon Quadrangle, Virginia, 7.5-minute series, 1980

We know little about the techniques employed by these surveyors, though, as we shall see, their work was remarkably accurate, easily related to modern landscape features. They probably relied on the main North American survey manual, John Love's *Geodesia* (1688), which went through thirteen editions. Love had worked in both Jamaica and North Carolina, and specifically directed his work toward North American surveyors. It would seem that the plane table and triangulation, helpful for the small clear plots of England, were not much used in the wilderness of South Carolina, where chain and compass were the rule. Curiously, it seems that there was little interchange between South Carolina and Jamaica, for of the nearly three hundred surveyors working in the two places at roughly the same time, only one name is common, and it is by no means certain that this John Henderson was one person.[29]

What is probably the earliest South Carolina estate map is reproduced as plate 4. By an unknown surveyor, it defines the boundaries of a grant of land "scittuate upon ye westermost side of ye westermost branch of Cooper River and belonging to the Honourable Colonal Joseph West, Governor and Landgrave of the Province of Carolina." The plat shows very little detail, for the land was as yet scarcely brought into cultivation. But the 3-inch border of fruits and flowers, including the native wild jasmine as well as pomegranates and strawberries, eloquently testifies to the decorative, celebratory quality of this type of map.

The later maps by Joseph Purcell, which are preserved in exceptionally large numbers, are not so elegant, but they are particularly effective in tracing the outlines of the eighteenth-century countryside. In 1785, for instance, Purcell drew "A Plan of Skinking Plantation belonging to Ralph Izard . . . situated on Santee River in Saint John's Parish" (figure 4. 9). The map is oriented eastward and is not easy to reproduce; I have therefore compiled an accompanying explanatory map (figure 4.10). From this we see that much of the land along the Santee River was low swamp land, sometimes "overflown at every frisk" (or flood). To the north of the river were indigo fields in the "high swamp," and to the south, behind the "negro houses," were further "fields of indigo," with a "settlement" in the middle of them. South again of this area were the "high pine land" and the "old fields, high land." Purcell indicated the escarpments and high ground with precise and ingenious shading, though this is not easy to see in the reproduction.

The distinctive curve in the river allows us to situate the site of the plantation very precisely on the U.S. Geological Survey map of 1919–20. As figure 4.11 shows, by this time virtually nothing was left on the ground of Skinking Plantation. The great lobe of marsh to the south of the curve in the river is plainly recognizable, as is the high ground upon which the main fields had stood; Purcell's shaded escarpments correspond closely with the high ground shown on the USGS map. But the only hint that there were once dwellings here is a road that mysteriously terminates. It might have been possible to detect the outlines of the plantation by archaeology, but it now lies beneath the waters of Lake Marion, drowned after the construction of the Santee Dam.

A little lower on the Santee River lay Richmond Plantation, belonging to Theodore Gaillard (1737–1805); having sided with the British, he had theoretically had his estates confiscated and been banished from the state. Purcell made a map of Richmond in about 1791 (figure 4.12).[30] It was a larger plantation that Skinking, and was devoted to

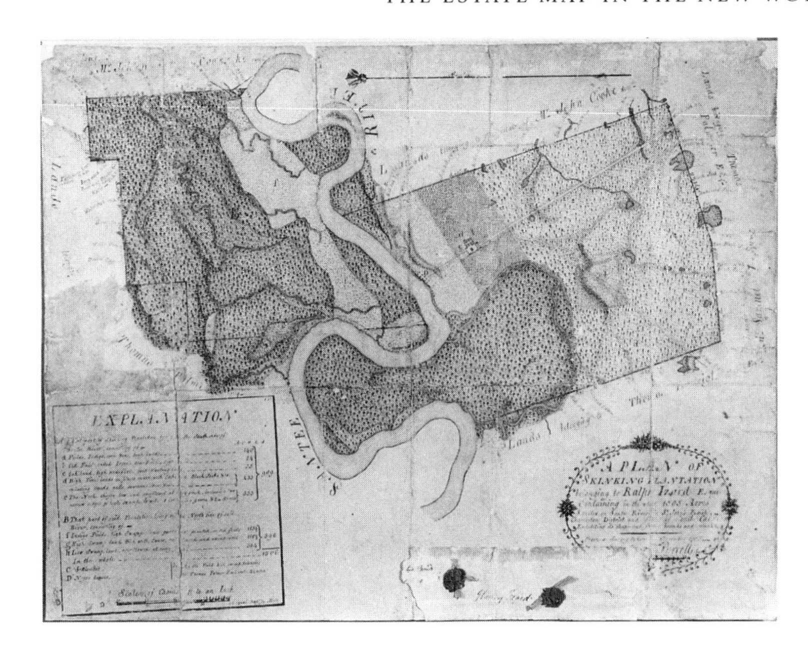

Fig. 4.9 Joseph Purcell, *A Plan of Skinking Plantation*, 1785 (South Carolina Historical Society)

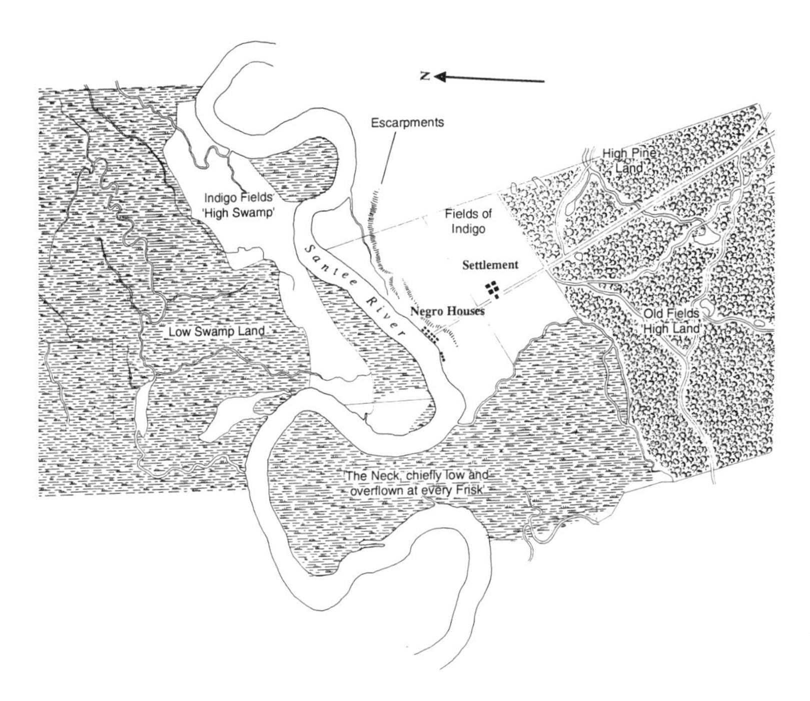

Fig. 4.10 Redrawing of Purcell's plan of Skinking Plantation

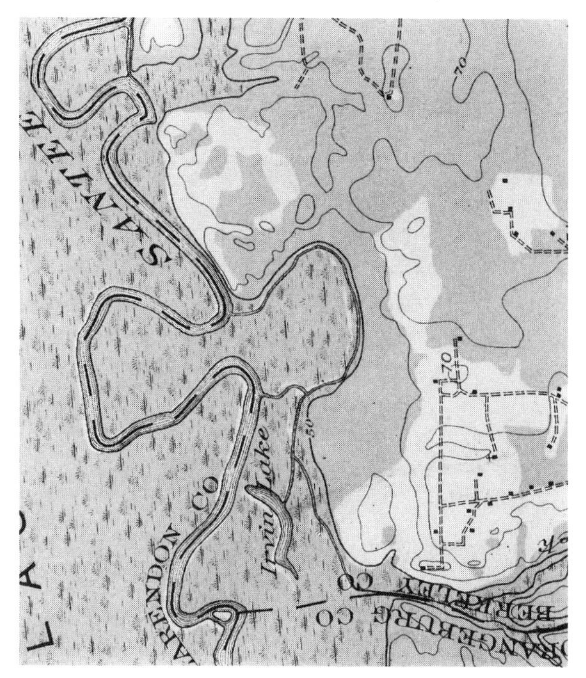

Fig. 4.11 Detail from the USGS Chicora Quadrangle, South Carolina, 7.5-minute series, 1919–20

the culture of rice, as the explanatory map shows (figure 4.13). A great marsh lies along the river, with "swamp land unimproved," and then come the five numbered rice fields, "banked and ditched." The living quarters for slaves and master are adjacent to the fields, but are on the high land behind the well-observed escarpment. East of "Richmond Settlement" is a crossroad, where the "public road from Murray Ferry to Kingstree" passes, and east again of that are the "high and flat land in woods," the "swamp land, unimproved in woods," and the "old field." Perhaps this old field had been an unirrigated rice field, in the time before planters progressed to the type known as the "controlled inland swamp" shown here.[31] The site of Richmond Plantation is at present ruinate, to judge by recent USGS maps (figure 4.14). The bend in the river stands out well, as does the escarpment, away to the north. Purcell's prominent crossroad is here marked "75," though there seems to be no sign of the buildings at "Richmond Settlement." About 3

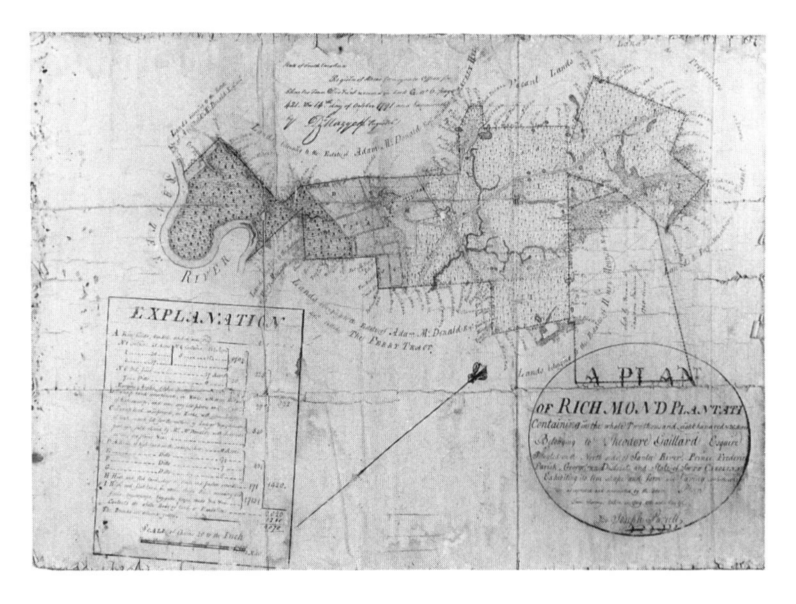

Fig. 4.12 Joseph Purcell, *A Plan of Richmond Plantation*, c. 1791 (South Carolina Historical Society)

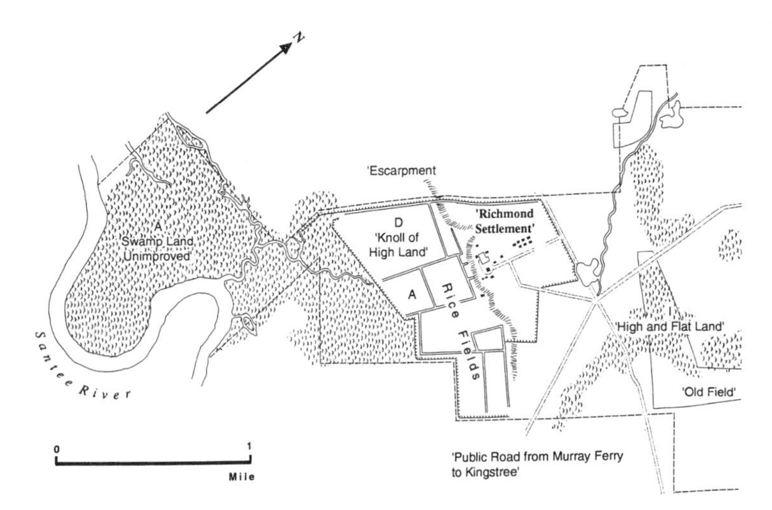

Fig. 4.13 Redrawing of Purcell's plan of Richmond Plantation

miles to the south, across the river, the USGS map (Pineville Quadrangle) marks Richmond Cemetery.

Six years later Purcell mapped Mount Pleasant and New Ground, plantations on the Upper Stono River (figure 4.15).[32] His map is reasonably clear, but its main features have been redrawn for greater legibility in figure 4.16. Mount Pleasant, on the left, had nearly 600 acres, and New Ground, on the right, nearly 300. The size of the slave quarters reflected this difference, for whereas twelve huts are shown for Mount Pleasant, only eight appear for New Ground. The rice fields are logically disposed in the valley bottom, irrigated (perhaps by tidal action) through the "Middle Drain." The "clear high land" above the valley bottom contains both fields and extensive "mixed woods"; here too are the main roads and the quarters for both masters and slaves. These plantations lay just southwest of Rantowles, on Route 17, but the "Middle Drain" is all that now appears to remain of them.

My final example of South Carolina estate maps comes from an area roughly 10 miles north of Beaufort, where Route 17 crosses the Combahee River (figure 4.17). Here in 1795 John Goddard surveyed a plantation known as Ferry Tract for Henry Middleton, creator of the magnificent gardens at Middleton Place on the Ashley River.[33] At Ferry

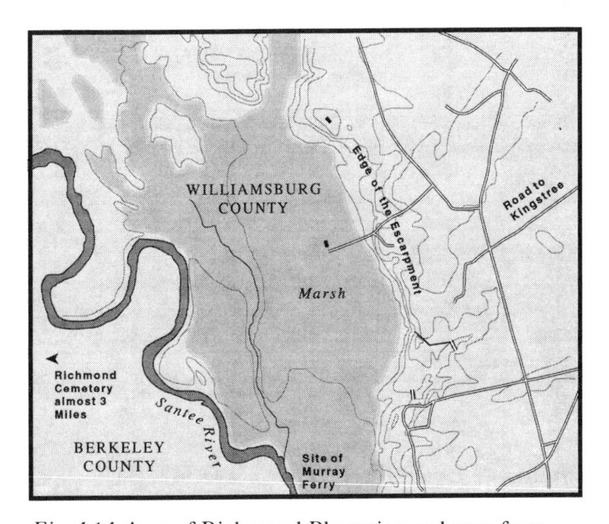

Fig. 4.14 Area of Richmond Plantation redrawn from the following USGS maps, South Carolina, 7.5-minute series: Butlers Bay, 1979; Greeleyville, 1990; Pineville, 1979; Saint Stephen, 1990

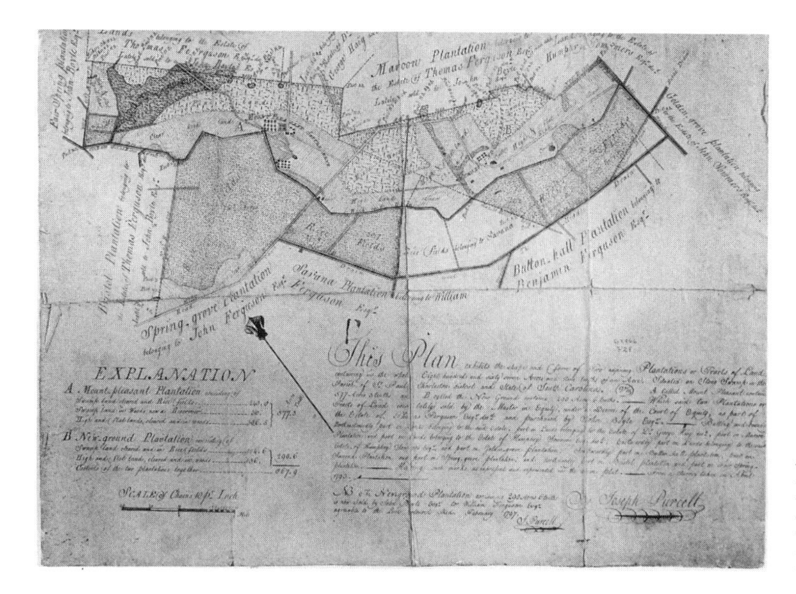

Fig. 4.15 Joseph Purcell, plan of Mount Pleasant and New Ground Plantations, 1797 (South Carolina Historical Society)

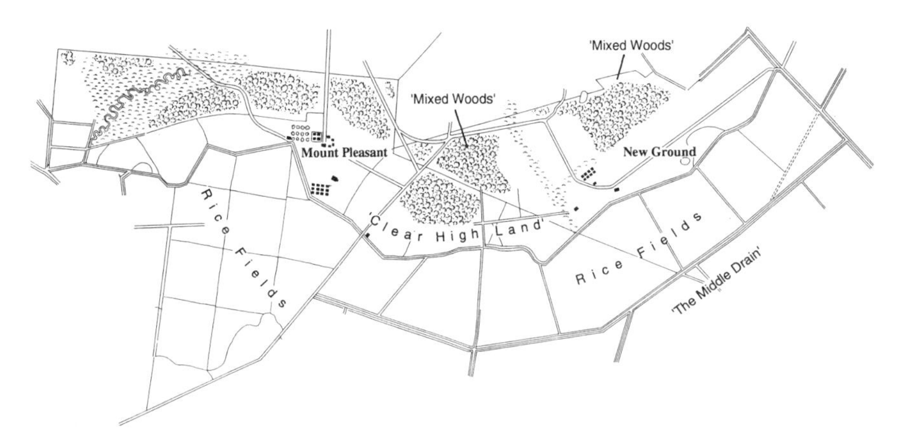

Fig. 4.16 Redrawing of Purcell's Mount Pleasant and New Ground plan

Tract were numerous "rice fields, tide land," and also the "old rice fields," described as being on "cleared inland swamp." Above this area were the pine lands; the estate was run from a substantial-looking great house, with nineteen slave houses nearby. The "Public Road to Combahee Ferry" ran through the southern part of this estate.

Much of this layout was still visible on the 1918 USGS Green Pond Quadrangle (figure 4.18). The Combahee Ferry was still the way to cross the river on the Old Savannah and Charleston Road, and the Ferry Tract buildings still appear much as they had in 1795; many of the field and road lines are also clearly recognizable. By 1988 much of this had changed (figure 4.19). Route 17 bridges the river, and the Ferry Tract buildings have largely disappeared. Many of the fields and roads, though, retain the same alignments as those of two hundred years

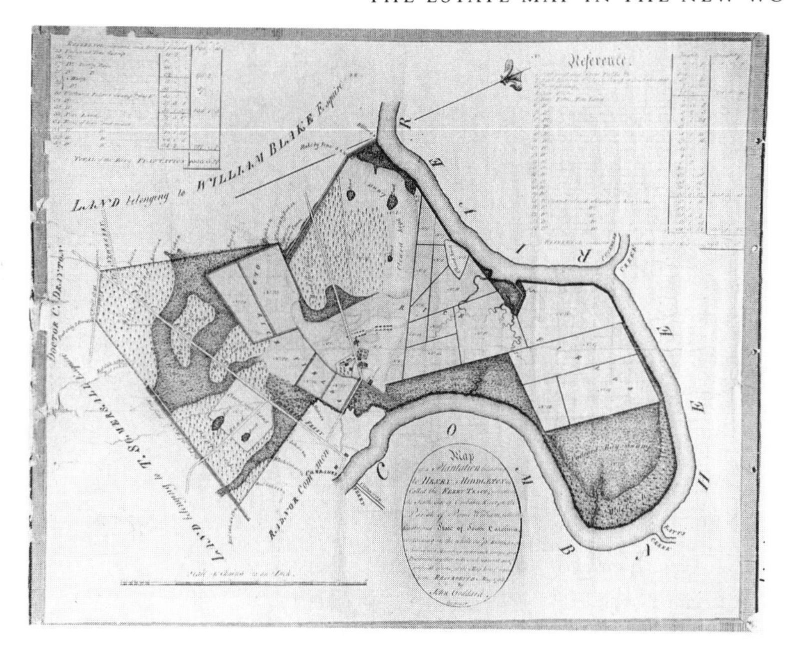

Fig. 4.17 John Goddard, *Map of a Plantation Belonging to Henry Middleton Called the Ferry Tract,* 1795 (South Carolina Historical Society)

earlier, the construction of the rice fields having marked the landscape almost indelibly.

On the whole, historians do not seem to have made as much use of these remarkable estate maps as they might have done. Henry A. M. Smith, it is true, used them extensively in his detailed topographical studies of the 1910s and 1920s.[34] But they have not been much used since then, though their extraordinary value is well shown in a recent article by William Lees.[35] Studying the development of Limerick Plantation, Lees uses Purcell's plans of 1786 and 1797, preserved at the South Caroliniana Library, to show the relative emphasis upon the "old rice fields" on the high ground, and the new ones on the tidal flats by the river. Curiously, although Lees relies so heavily upon these plans, he does not reproduce either of them in his very interesting article. Perhaps this is because they are not easy to reproduce, as we have seen. But it is not difficult to make accurate drawings from them, and such drawings ought to be used, not only for studies

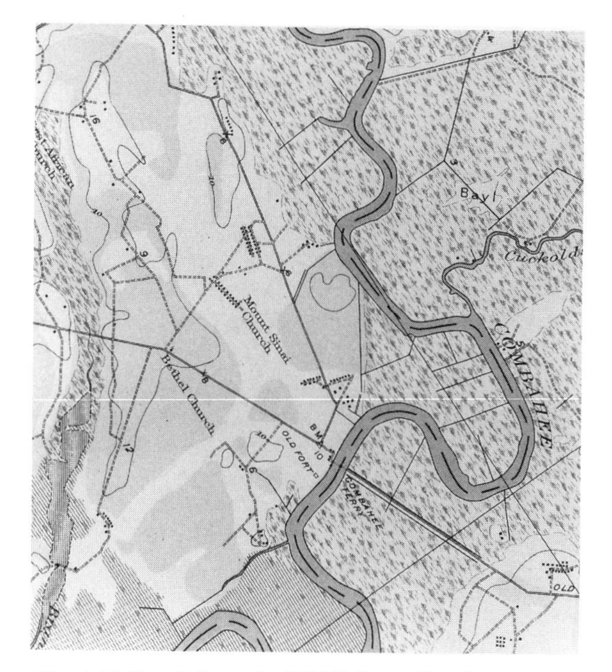

Fig. 4.18 Detail from the USGS Green Pond Quadrangle, South Carolina, 15-minute series, 1918

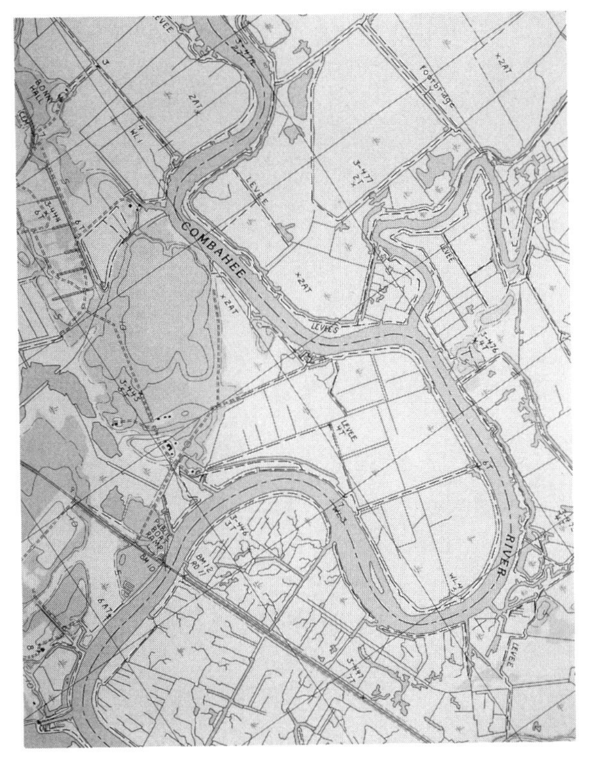

Fig. 4.19 Detail from the USGS White Hall Quadrangle, South Carolina, 7.5-minute series, 1988

in economic history, but also for work in the increasingly important area of landscape history.

Georgia and Florida

To the south of South Carolina, Georgia had been established in 1733, as a colony without slaves.[36] From the 1740s, however, it began following the example of the area to the north; slaves were imported, and rice plantations began to develop on the four main rivers, the Savannah, Ogeechee, Altamaha, and Satilla. The development was much as in South Carolina, with inland rice swamps giving way toward the end of the eighteenth century to tidewater cultivation; however, Georgia never attained the same level of production as its neighbor to the north, and her planters do not seem to have commissioned anything like the same number of estate maps.

Some, however, were compiled, mostly in the early nineteenth century. In *Savannah River Plantations* Mary Granger offers redrawings of estate maps of 1819 and 1830, as well as later ones; they seem to resemble closely the maps of South Carolina.[37] Working southward from Savannah, we come to the Ogeechee River; here in 1769 Samuel Savery drew a fine plan of James Read's estate.[38] This plan is reproduced as plate 6; figure 4.20 shows the same area from a USGS map of 1976. It lies about 10 miles out of Savannah, to the south of Route 204, here shown as Fort Argyle Road. Fort Stewart Military Reservation abuts it to the south, and the area has not greatly prospered; none of the early property lines has survived. In comparing the two maps, we are struck by the skill with which Savery drew the Ogeechee River, down to the small tributary up by Spring Hill Cemetery.

Sporadic evidence of estate maps survives from plantations farther to the south. In 1929, in his *Life and Labor in the Old South,* Ulrich Phillips published *Map of Hopeton, 1821,* taken from the Hopeton crop record book at the University of North Carolina Library. This plantation lay near the mouth of the Altamaha River and belonged at the time to James Cooper. The anonymous cartographer showed the estate in the usual way, taking care to show fields as producing not only rice, but also cotton, corn, "pease," and potatoes. Finally, at least one estate plan survives from Florida, though that state was hardly settled by Europeans (apart from the long-standing Spanish enclave at St. Augustine) before the middle of the nineteenth century. The map of Beauclerk Bluffs, on the St. Johns River, was drawn in 1771 and is now preserved at the Public Record

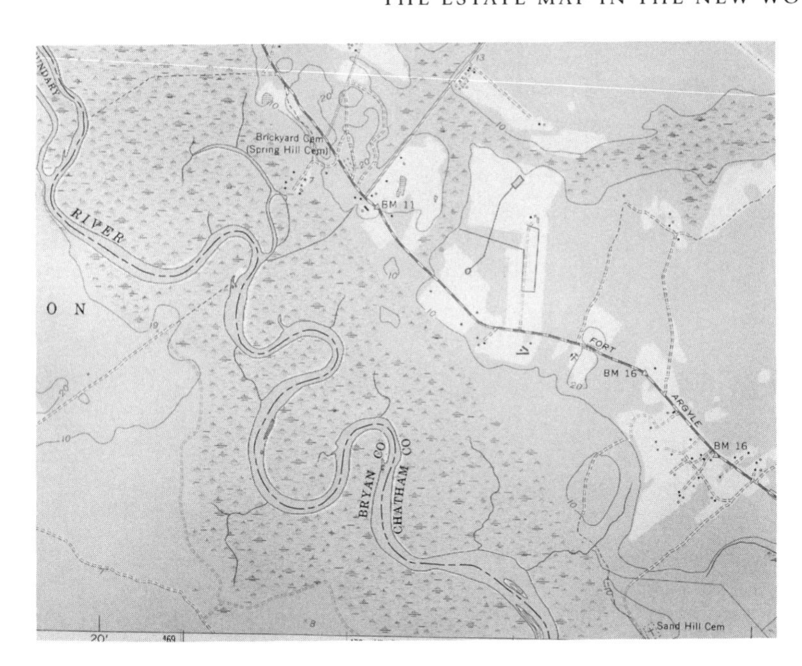

Fig. 4.20 Detail from the USGS Meldrim SE Quadrangle, Georgia, 7.5-minute series, 1976

Office in London.[39] The surveyor distinguishes the different crops and includes a sketch of one of the slave quarters.

The Lower Mississippi River

The lower Mississippi River region saw the steady expansion of European occupation, from the time of John Law in the early eighteenth century onward, and very substantial estates developed along the Mississippi River. However, the emergence of estate plans was late and sporadic. They are now preserved in a variety of places; the City Hall Annex in New Orleans has some late-nineteenth-century plans,[40] the Historic New Orleans Collection has others,[41] some are in the Louisiana State Museum in New Orleans,[42] and at least one is in the Southern Historical Collection at the University of North Carolina.[43]

The finest of these plans is undoubtedly the *Plan de l'habitation de feu J. B. de Marigny . . . ,*" preserved at the Historic New Orleans Collection.[44] It shows a characteristic French-inspired long lot, stretching away at right angles from the Mississippi River into the bush. As was normally the case with Louisiana plantations, the buildings were grouped by the river, followed by the fields (usually growing sugarcane), which shaded off into the bush. The Marigny Plantation now forms a substantial slice of metropolitan New Orleans, east of Elysian Fields Avenue, where its outline may still be readily traced on the modern street plan.

Perhaps more characteristic, in its utilitarian appearance, is the map of Houmas Plantations preserved at the Louisiana State Museum. Figure 4.21 is a redrawing of this map, originally compiled about 1880. The seven plantations are grouped in a fan shape on a bend in the Mississippi River, with their fields running back into the bush at the north. Substantial roads connect the estates with each other and with the river road, and the central buildings are shown in detail. It is easy to pick out the

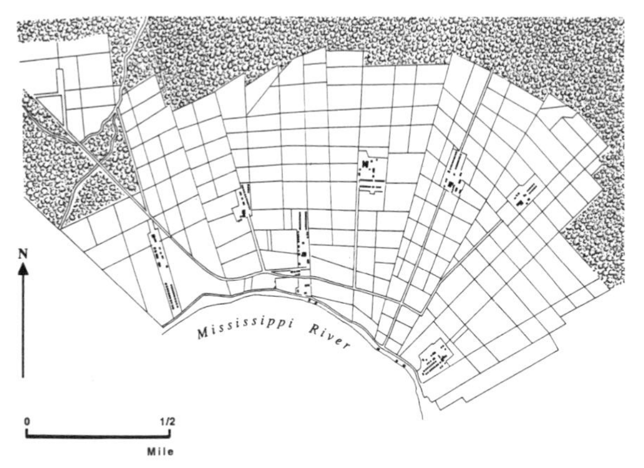

Fig. 4.21 Redrawing of the anonymous map of Houmas Plantations, c. 1880 (Louisiana State Museum, New Orleans)

main structures, the sugar mill, distillery, and so forth; the former slave quarters also show up well. A map like this allows very telling comparisons to be made with the present-day terrain, for the line of the fields and roads has changed very little.[45]

Estate maps can sometimes be found for other sites, farther north on the river. A particularly fine one was recently acquired by the Mississippi Department of Archives and History; the *Plan of Concord, the Plantation, and Residence of the Late Major Stephen Minor* was executed by George Dougherty in 1829.[46] Major Minor came from Virginia, where he may have seen maps of this kind; Concord was an important plantation containing "the outstanding Spanish-era house in the Natchez region." Perhaps there were not many other plantations of equal significance, to entice their owners to record them in this way.

Other Regions of North and Central America

The large part of the United States that was formerly ruled from Mexico City had cartographic traditions of its own. The vast collection of maps in the Archivo de la Nación in Mexico City contains many plans of rural estates, some of which are not unlike the estate maps that we have been studying.[47] Unfortunately, very few studies have used this remarkable material, which derives from both Spanish and Amerindian traditions.[48] Even a cursory examination of the catalog shows, though, that the Mexican archives contain detailed topographical maps ranging from the late sixteenth to the late eighteenth century; the later ones obey the rules concerning planimetric delineation that were common in eighteenth-century Europe.[49]

In California, where the Spanish/Mexican presence became considerable during the eighteenth century, a form of estate plan known as the *diseño* eventually became common; indeed, such maps were required by Mexican governors before land could be registered.[50] The *diseños* almost always have scale and orientation, and some indication of the main roads, watercourses, and buildings of the ranches that they mostly represent. Often they show vast areas of relatively infertile land, and so have no need to go into the precise description of individual fields.

The other nonanglophone area of North America that developed its own system of topographical mapping was the St. Lawrence valley. Here from the middle of the seventeenth century onward the French colonists were establishing themselves on their distinctive long lots, surveyed as early as 1641 by Jean Bourdon. Early in the eighteenth century the area was mapped in three sheets, covering the regions of Quebec, Trois-Rivières, and Montreal, by Gédéon de Catalogne.[51] This map was large-scale enough to show individual holdings, but not detailed enough to show the buildings on them. Some of the seigneuries of the St. Lawrence Valley, particularly the ecclesiastical ones, were quite extensive, but they do not seem to have generated what we have defined as estate maps, perhaps because on most of them the agriculture was neither very intensive nor very market-oriented.

Some of these considerations apply to one final type of topographical map, the "village views" produced by the Shakers.[52] These naive but often very informative maps of Shaker communities are found for many different states, including Indiana, Maine, Massachusetts, and New York. Some commentators have regarded them as somewhat incongruous products for a group that prided itself on its unostentatious modesty. But there is no denying that these village views, some of which closely resemble European estate maps, performed something of a celebratory function: see, the artist says, how neat and well-ordered our communities are!

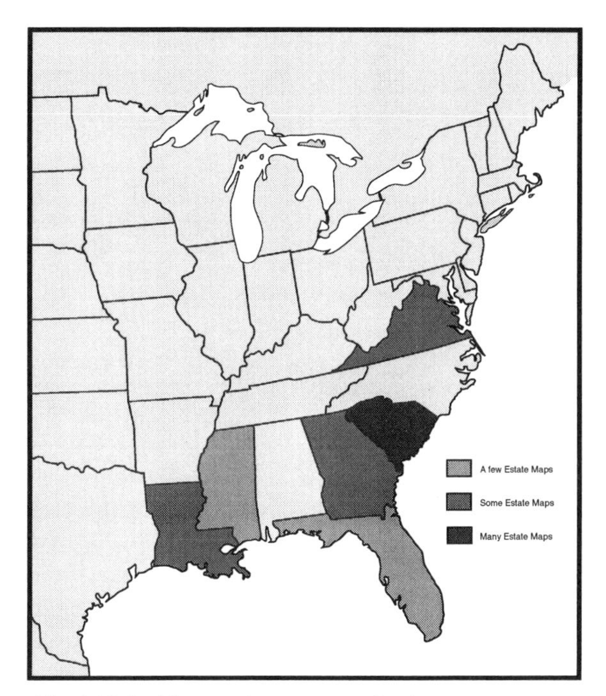

Fig. 4.22 Incidence of estate maps in the eastern United States

Conclusion

In the New World as in the Old, the incidence of estate maps tells us much about social and economic conditions. In this respect, it is remarkably instructive to examine the maps in S. B. Hilliard's *Atlas of Antebellum Southern Agriculture.* Of the many possible variables, including land types, population, farm size, and crop types, the one that coincides most closely with the incidence of estate maps—in a very striking way—is the proportion of slaves to the total population. The maps showing census data of this kind from 1790, 1800, 1810, 1820, 1830, 1840, and 1850 might almost be delineating the same information as the map showing the distribution of estate maps (figure 4.22). In a way this is not surprising, for, as we have seen, the estate plan was in the New World originally the product of slave societies. It is interesting to consider that Virginia, which had a relatively low proportion of slave to free inhabitants, also had a rather low incidence of estate plans. By the 1860s the Mississippi valley was becoming an area of high slave/free ratios, and we might have expected a greater proliferation there of estate plans. But the plantations of the valley were almost predictable in their shape, with the great house down by the river, the quarters and works behind them, and the fields stretching into the bush. Perhaps owners felt no need to delineate this well-understood layout; perhaps too, as we have seen in Europe, there was less need to draw estate plans of an area that by then had quite satisfactory large-scale general maps.

In conclusion, it might be helpful to refine our understanding of the nature of the estate map in the New World, by comparing it with a map type that was in many respects its antithesis—the nineteenth-century North American county atlas map.[53] Such maps were not commissioned by an individual proprietor, but were the work of a cartographic entrepreneur, who then sold them to the individual farmers. They were not manuscript, but printed in quite large numbers in some center like Philadelphia or Chicago; they could consequently be sold far more reasonably than could the specially commissioned estate map.

On the other hand, the two map forms had some points in common, apart from representing individual agricultural units. The county atlas maps could thrive only in areas of prosperous agriculture, like the estate maps. Both map types eventually formed part of a genre that can be traced historically over considerable areas of country. Both, moreover, shared a celebratory role, which sometimes induced surveyors and cartographers to make their clients' properties seem somewhat grander than they were. For Everyman Farmer in the Midwest, as for tidewater planters, it was important to have a map that made the most of his possessions, and this was one of the leading characteristics both of the county atlas maps and of the colonial estate map, with its lineage stretching back to the proud squires of the old countries.

Notes

1. See particularly Stephen Saunders Webb, *The Governors-General: The English Army and the Definition of Empire, 1569–1681* (Chapel Hill, N.C., 1979).

2. Jack Greene, "Colonial South Carolina and the Caribbean Connection," *South Carolina Historical Magazine* 88 (1987):192–210.

3. See for instance the estate plan of 300 acres of land near Holetown, Barbados, surveyed by John Hapcott, 10 October 1646, recently acquired by the John Carter Brown Library (Providence, R.I.).

4. See M. J. Chandler, *A Guide to Records in Barbados* (Oxford, 1965).

5. *Saint Vincent,* 1:50,000, Directorate of Overseas Surveys, 1968.

6. See the bibliography in Gabriel Debien, *Les Esclaves aux Antilles françaises, XVIIe–XVIIIe siècles* (Basse-Terre, Guadeloupe, 1974), and Debien, *La Sucrerie Galbaud du Fort, 1690–1802* (n.p., 1941), where a plan is reproduced. Debien once remarked to me that he had observed many estate maps in the country houses of French families with West Indian connections.

7. Bibliothèque Nationale, Paris, MS fr. 6256, "Registre des affaires domestiques de M. Turgot, gouverneur de Cayenne et de la Guyane"; the plan follows a section of accounts.

8. Archives Nationales, Paris, Saint Domingue, N III, 17.

9. For the plan, see Archives Nationales, Saint Domingue, N 2. The papers for this plantation are in the Public Record Office, London; see Debien, *Les Esclaves,* 34.

10. On the *pinturas,* or maps, that accompanied many of the *relaciones geográficas* commissioned by the Spanish crown, see Barbara Mundy, "The Maps of the *Relaciones Geográficas* of New Spain, 1579-c. 1584," Ph.D. thesis, Yale University, 1993.

11. Described by Marlin D. Clausner in *Rural Santo Domingo: Settled, Unsettled, and Resettled* (Philadelphia, 1973), and by Duvon C. Corbitt, "Mercedes and Realengos: A Survey of the Public Land System in Cuba," *Hispanic American Historical Review* 19 (1939):262–85.

12. See Johannes Hartog, *Curaçao: From Colonial Dependence to Autonomy* (Aruba, 1968).

13. See Daniel Hopkins, "An Extraordinary Eighteenth-Century Map of the Danish Sugar-Plantation Island St. Croix," *Imago Mundi* 41 (1989):44–58; Hopkins, "Jens Michelsen Beck's Map of a Danish West Indian Sugar-Plantation Island," *Terrae Incognitae* 25 (1993):99–114.

14. See the excellent color reproduction in Susan Danforth, ed., *The Land of Norumbega: Maine in the Age of Exploration and Settlement* (Portland, 1988).

15. Peter Benes, *New England Prospect: A Loan Exhibition of Maps at the Currier Gallery of Art* (Boston, c. 1981).

16. I here acknowledge the great assistance afforded me by Jo Mano of the State University of New York, New Paltz, who kindly searched the New York State Archives for me. The Maryland Historical Society repository does contain the interesting printed *Plan of Carrollton Hall,* drawn by W. M. Dawson, Jr., about 1850.

17. See Fairfax Harrison, *Landmarks of Old Prince William* (Richmond, 1924), which has an appendix, "Maps and Map-Makers," 2:601–51.

18. For this information I am much indebted to Allen Meyer, of the Chicago Map Society, who made a search in the Virginia State Library and Archives.

19. See for instance James Minge's 1701 plan of William Byrd's Westover Plantation, reproduced in Roger Kain and Elizabeth Baigent, *The Cadastral Map in the Service of the State: A History of Property Mapping* (Chicago: 1992), 272.

20. Sarah S. Hughes, *Surveying and Statesmen: Land Measuring in Colonial Virginia* (Richmond, 1979), 47.

21. Ibid., 48.

22. Library of Congress; see Lawrence Martin, ed., *The George Washington Atlas* (Washington, 1932); Donald Wise, "The Young Washington as a Surveyor," in *Northern Virginia Heritage* (1979).

23. See the plats in *Thomas Jefferson's Farm Book,* ed. Edwin Morris Betts (Princeton, 1953).

24. See Linda-Marie Pett-Conklin, "Cadastral Surveying in Colonial South Carolina: A Historical Geography," Ph.D. thesis, Louisiana State University, 1986.

25. See S. B. Hilliard, "Antebellum Tidewater Rice Culture in South Carolina and Georgia," in *European Settlement and Development in North America: Essays on Geographical Change in Honor and Memory of Andrew Hill Clark,* ed. James R. Gibson (Toronto, 1978). For an understanding of low-county plantations, I have largely relied on Samuel Gaillard Stoney's classic *Plantations of the South Carolina Low Country* (Charleston, 1938; reprint, New York, 1989).

26. Many estate plans are in the H. A. M. Smith Collection, joined now by the Balzano and Gaillard Collections (see *Carologue: A Publication of the South Carolina Historical Society*) [autumn 1992 and winter 1993]). I would like to express my gratitude to successive curators in the Fireproof Building (especially Cam Alexander and Mary Giles) for their great help to me. I have also received good advice from Chuck Lesser, of the South Carolina Department of Archives, and from Frederick Holder.

27. Pett-Conklin, "Cadastral Surveying," 211–45.

28. The surveyors' names are Peter Belin, Nathaniel Bradwell, John Diamond, John Fenwick, Charles and Issac Gaillard, John Goddard, John Hardwick, William Hemingway, Ephraim Mitchell, Joseph Purcell, Henry Ravenel, Samuel Wells, and John Wilson.

29. This comparison was established by using the lists provided by Pett-Conklin and B. W. Higman, *Jamaica Surveyed: Plantation Maps and Plans of the Eighteenth and Nineteenth Centuries* (Kingston, 1988).

30. This map is reproduced in the article on large-scale maps by Louis DeVorsey in *From Sea Charts to Satellite Images: Interpreting North American History through Maps,* ed. David Buissesret (Chicago, 1990), 80–81.

31. See Hilliard, "Antebellum Tidewater Rice Culture."

32. An analysis of the plans of these and other plantations is offered by Philip Morgan in "The Development of Slave Culture in Eighteenth-Century Plantation America," Ph.D. thesis, University College, London, 1977.

33. On Middleton Place, see Stoney, *Plantations of the South Carolina Low Country.*

34. Published in the *South Carolina Historical Magazine* between 1917 and 1928, and then gathered into

three volumes published at Spartanburg in 1988.

35. William Lees, "The Historical Development of Limerick Plantation," *South Carolina Historical Magazine* 82 (1981):44–62.

36. See Douglas Wilms, "The Development of Rice Culture in Eighteenth-Century Georgia," *Southeastern Geographer* 12 (1972):45–57.

37. Mary Granger, ed., *Savannah River Plantations* (Savannah, 1947).

38. Meldrim SE Quadrangle, Georgia, 7.5-minute series. The Savery map was reproduced by the National Archives as a poster for the exhibit "Taking the Measure of the Land," c. 1971.

39. See George Kish, *The Discovery and Settlement of North America, 1500–1865: A Cartographic Perspective* (New York, 1978), 46.

40. For instance, of Barataria (1866), City Hall Annex, New Orleans, book 65/72, and of Nairn (1884), book 100/2.

41. I want to acknowledge the help of the late John Mahé, in using the Historic New Orleans Collection.

42. Rose Lambert was my guide to the Louisiana State Museum collection, and Bernard Lemann to the New Orleans collections in general.

43. *Topographical Map of Magnolia Plantation (Plaquemines) 1880*, Southern Historical Collection.

44. Reproduced in David Buisseret, ed., *Rural Images: The Estate Plan in the New and Old Worlds* (Chicago, 1988), 29.

45. See the eighteenth-century map and the corresponding satellite image in Buisseret, *From Sea Charts to Satellite Images*, color plates 1, 2.

46. My attention was drawn to this map by Patricia Galloway of the Mississippi Department of Archives and History.

47. See the *Catálogo de ilustraciones*, 14 vols. (Mexico City, 1979–82).

48. One of the few is Gisela von Wobeser, *La formación de la hacienda en la epoca colonial* (Mexico City, 1983).

49. See for instance Buisseret, *Rural Images*, 36.

50. On this subject see Robert H. Becker, ed., *Designs on the Land: Diseños of California Ranches* (San Francisco, 1969).

51. Reproduced with a bibliography in David Buisseret, ed., *Mapping the French Empire in North America* (Chicago, 1991).

52. See Robert P. Emlen, *Shaker Village Views* (Hanover, N.H., 1987).

53. See Michael P. Conzen, "The Country Landownership Map in America," *Imago Mundi* 36 (1984):9–31.

Chapter FIVE

THE MAKING *of* JAMAICAN ESTATE MAPS *in the* EIGHTEENTH *and* NINETEENTH CENTURIES

B. W. HIGMAN

The Surveyor's Role

During the eighteenth and nineteenth centuries the fortunes of the Jamaican surveying profession depended very much on those of the plantocracy. Surveyors and planmakers charted the expansion and refinement of the plantation system under slavery, and created a vital record of the transformation of land-use patterns following emancipation. A substantial sample of their work survives in the plan collection of the National Library of Jamaica and supplies the raw material for this and chapter 6. The surveyors' role was not merely documentary; they injected their own ideas and attitudes into the plans they produced, taking an active part in the transformation of the island's spatial structure and the ordering of society. To understand the maps they produced, we need to study the legal framework within which they operated, their numbers, their social origins, their education and training, and their professional and business organization, as well as the techniques employed in their work.

The Legal Framework

From the English conquest of 1655 to the abolition of the Legislative Assembly after 1865, Jamaican law relating to land and land surveyors was controlled by the plantocracy.[1] The Act for Regulating Surveyors of 1683 provided that the surveyor general should be responsible for survey only when the Crown was a party; otherwise it permitted "any person or persons whatsoever to survey, re-survey, and run any dividing lines, and give plats of any land."[2] The weakness of this provision soon became manifest, and later in 1683 the assembly passed an Act for Further Directing and Regulating the Proceedings of Surveys, by which the activities of surveyors were much more closely defined; this did not prevent them from often being regarded as scoundrels, ready to take advantage of their office.

These acts of 1683 provided the basis upon which the early surveyors operated. In 1780 a further act regulated their "exorbitant charges," and this seems to be the first Jamaican law to make reference to surveys of the internal layout of plantations, rather than mere boundary-line running. The 1780 act also included a clause seeking to establish basic qualifications for surveyors. It noted that "many disputes and tedious expensive law-suits have arisen, . . . through the incapacity and ignorance of persons presuming to act as surveyors of land." Thereafter, no person was to be "appointed to act as surveyor of land" in Jamaica, "until he hath duly served some sworn surveyor as an apprentice, for at least five years in this island, or hath been a sworn surveyor in Great Britain or Ireland, and hath undergone an examination by and before three surveyors of land, to be nominated by the Supreme Court of Judicature of this island, as his qualification." This act and the scale of fees that is established governed surveyors until the time of emancipation.[3] In 1858 the land surveyors' act was promulgated; as in 1780, apprenticeship for five years, followed by examination, was the main route to a commission.[4] The examination was now to be public and to "embrace a knowledge of mathematics, theoretical and practical land surveying, including trigonometrical and railway surveying, drafting, plotting, protracting and isometrical and topographical drawing." Concerning the making of plats, the act of 1858 followed almost word for word the provisions of 1683.

The introduction of crown colony government was soon followed by a new law relating to land surveyors, in 1869.[5] There were some minor modifications, and the subjects examined were expanded to include "arithmetic, algebra as far as simple equation, plane geometry, plane trigonometry, topographical drawing, the practical use of the principal instruments used in surveying, and their adjustments, and the theory and practice of land and railway surveying." The final land surveyors' law of the nineteenth century was enacted in 1894, but none of the estate maps discussed here was produced under its provisions.

The development of Jamaican legislation relating to surveyors between 1683 and 1894 followed fairly closely the broad trends of British enactments,[6] but the demands of the plantation system and of colonization created differences in emphasis. Only gradually did the legislators begin to show some interest in the surveying of the internal layout of the plantations, and then their concern was confined largely to the question of fees. But the emergence of this interest about 1780 did match quite closely the beginnings of the golden age of the Jamaican plantation surveyor, and it was in the same period that the laws began to seek to control surveyors' qualifications.

The Surveyors and Their Professional Organization

All surveyors practicing in Jamaica were required by the act of 1683 to hold commissions, but the lists are defective before 1780, between 1847 and 1860, and between 1868 and 1881. Only eight "surveyors in commission" were listed in 1780, and it is unlikely that they were more numerous in former times. Their number expanded rapidly to reach a peak of thirty-four in 1795 (figure 5.1), but by 1810 a period of decline set in, lasting until about 1865. These trends followed the fortunes of the plantation economy, the sugar and coffee industries reaching their maximum geographical spread and output between 1790 and 1810. The late-nineteenth-century revival in the profession depended on government employment rather than demand from the plantation system, and the peak of 1795 was not exceeded until 1910.

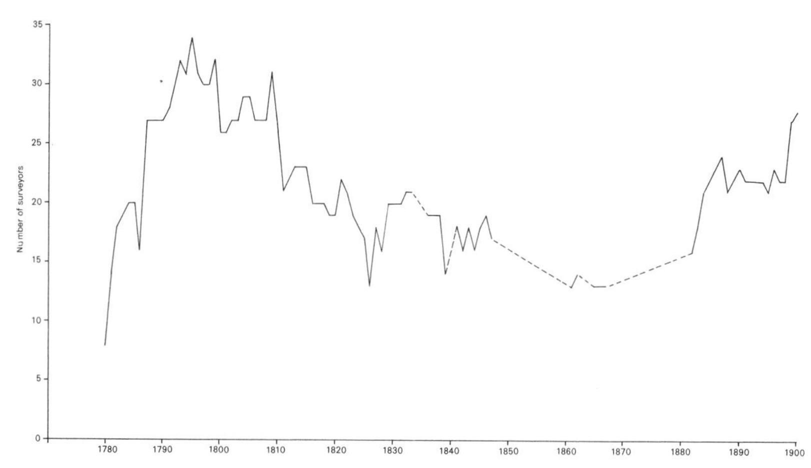

Fig. 5.1 Surveyors in commission in Jamaica, 1780–1900

To a certain extent, it is possible to identify the wider body of land surveyors, including assistants and apprentices, by using the collection of plans in the National Library of Jamaica, together with the lists published in the *Jamaica Almanack* and *Handbook of Jamaica*. A total of 220 individual surveyors are known by name, but for only a small proportion of these are further details available. This means that identification is not easy, particularly in a profession that often passed from father to son with the same personal names.

The total number of commissioned surveyors working in Jamaica between 1700 and 1900 was about 300.[7] Of the 220 persons identified by name, dates in commission are unknown for 40, but only 12 of these produced plans in periods for which lists are available; a small minority of plans were the work of surveyors not known to have been commissioned. Thus almost two-thirds of the total population of surveyors is known, and this proportion compares favorably with that achieved by Peter Eden's *Dictionary of Land Surveyors and Local Cartographers of Great Britain and Ireland, 1550–1850*.

It is more difficult to assess whether the number of surveyors working in Jamaica should be regarded as a relatively large or small body. In Scotland, for example, the peak period of land-surveying activity around 1815 employed more than ninety full-time surveyors in any one year, or about one surveyor per eighteen thousand people.[8] In Jamaica, at the peak date of 1800, the ratio was as high as one surveyor in commission for every twelve thousand people. Allowance is not made here for the number of surveyors in Jamaica working only part-time, but the ratio does include the great majority who were slaves, so it seems fair to conclude that Jamaica did in fact support a relatively large population of surveyors.

No corporate body emerged for the surveying profession in Jamaica during the eighteenth and nineteenth centuries, and no Jamaican surveyor seems to have been a member of the British Royal Institution of Chartered Surveyors, established in 1868, though the institution trained many colonial land surveyors in the late nineteenth century.[9] John Mann had been granted the title of surveyor general in 1661, but nothing is heard of the office during the eighteenth century, and the first reference to a crown surveyor appears in 1811, when the post was held by John Fullarton.[10]

In 1837 three crown surveyors were appointed, one for each of the counties, in order to supervise the massive demand for land following emancipation. They were appointed by the governor under a commission and "employed in the public work of their respective counties, making surveys of parish or county boundaries, any crown lands, or lands about forts or fortifications," but received no stipend from the government.[11] This system remained in force until 1865, but disappeared early in the administration of crown colony government, when Sir John Peter Grant created a Survey and Lands Division within the Public Works Department.

This division was headed by Thomas Harrison, who had been crown surveyor for Surrey since about 1852. It was said in 1883 that "the necessity for such a Department had long been felt, for there was no officer before the appointment of the Government Surveyor whose special duty it was to look after the lands belonging to Government, and these were scattered about in every part of the island, most of them neglected and many unknown."[12] Harrison held the post until 1892, and in 1894 it was given to Colin Liddell, Harrison's son-in-law, who remained in charge until his death in 1916.[13] Apart from this brief flurry of government activity after 1865, however, land surveying in Jamaica was chiefly a branch of private enterprise, controlled only by law and custom.

Social Origins and Wealth

The social origins and wealth of Jamaica's surveyors are hazy, though it is at least certain that they were all males, and most were thirty-five to fifty years of age.[14] Otherwise the censuses are silent; there is no positive evidence that colored or black men entered the profession, but it is possible that some had done so by the late nineteenth century. In the early days of English colonization all of the island's surveyors were immigrants, trained in English techniques and practices. British-born surveyors continued to come to work in the island throughout the period, but the impression is strong that in the years of greatest activity, between 1780 and 1850, Jamaican-born creoles dominated the profession. The system of apprenticeship and the hereditary character of the occupation contributed to this tendency, and it was also served by the failure of most surveyors to attain sufficient wealth to acquire plantations or become absentees. They were generally considered "professional" men, in the nineteenth century at least, but the early image of the scoundrel was hard to shake off.[15]

In the late eighteenth century the fee for a surveyor's commission stood at £100; this made it the highest of the fees imposed by the governor's secretary, the next largest being £50 for a chief justice of the Grand Court. The surveyor's fee was increased to £150 in 1838 ,[16] so that the high cost of obtaining a commission, together with the educational requirements, confined entry to the profession to the middle rank of moderately wealthy whites. Only when surveyors became scarce was the fee for a commission reduced, to £60 in 1858 and £30 in 1894.[17]

The fortunes of individual surveyors varied widely, depending on many factors beyond their ability or success as practitioners. Inheritance, marriage, and speculation often provided the base for a larger fortune and entry into the plantocracy. The best evidence of relative wealth is provided by the inventories of personal property taken at death for the assessment of a hereditaments tax. These cover the period from 1674 to 1881; the earliest of thirteen known inventories is that of John Rome the elder, who died in 1797.

Analysis of these thirteen inventories shows considerable variation, from relative penury to substantial wealth.[18] The mean value of personal property was about £10,000, but only four surveyors exceeded this amount. Only those who established themselves as planters became really wealthy, and most surveyors' incomes were only middling for Jamaica whites. During the period of slavery all of the surveyors owned at least five slaves, most of them employed as domestics, but only those who possessed plantations had more than twenty; this pattern placed surveyors in the middle rank of Jamaican slave owners.[19] The number of examples is too few for certainty, but the impression exists that the surveyors' fortunes reached a peak about 1820 and then declined, following trends in the plantation economy. Few became absentees; many worked in partnerships, and father-son succession was common. Some worked part-time, and many served in the militia, the chief arm of white power in Jamaica. Some held sway in the courts or represented their fellows in vestry or assembly.[20]

Education and Training

Evidence regarding the education and training of Jamaica's land surveyors is less plentiful, and must be sought in materials other than inventories and lists of public offices. Some of the best information of this kind is provided by the record of statements made before a committee of the assembly appointed to prepare a bill to regulate the work of surveyors in 1842. Three surveyors were examined by the committee, and their careers throw a good deal of light on questions of education and training.

The replies of Edward McGeachy, John Matthias Smith (1799–1854), and Richard Wilson, together with what we know about Thomas Harrison (c. 1823–94), allow us to identify two general features about the careers of Jamaica's surveyors.[21] These are that schools of surveyors perpetuated practices through father-son succession, apprenticeship, and partnership, and that spheres of surveying activity were usually confined to small regions of the island.

The lines of descent of Jamaican surveying "schools" can be reconstituted only partially, since there are many gaps in the knowledge of kinship and apprenticeship relations. One of these networks covered more than a century, from the firm of Gordon and Grant in the 1770s to Thomas Harrison and Colin Liddell.[22] The surveyors in this school practiced only in the eastern half of the island, though they had spread as far west as Manchester by the later nineteenth century. Only in a small number of cases can the master-apprentice relationship be confirmed, but the changing composition of the surveying partnerships is strongly indicative of the line of descent.

A second network encompassed only sixty years, from 1775 to 1831.[23] This school had its origins in the east, but most of its later members worked in the developing western and northern parishes. Individual surveyors, however, generally practiced in no more than two or three contiguous parishes over their careers. Some examples of these regional patterns in the peak period of the 1790s are provided in figure 5.2. During the nineteenth century, as the number of surveyors declined, individuals tended to range more widely, though the broad division between the east and the west of the island remained.

Jamaica's experience in the development of schools of surveyors through succession and the

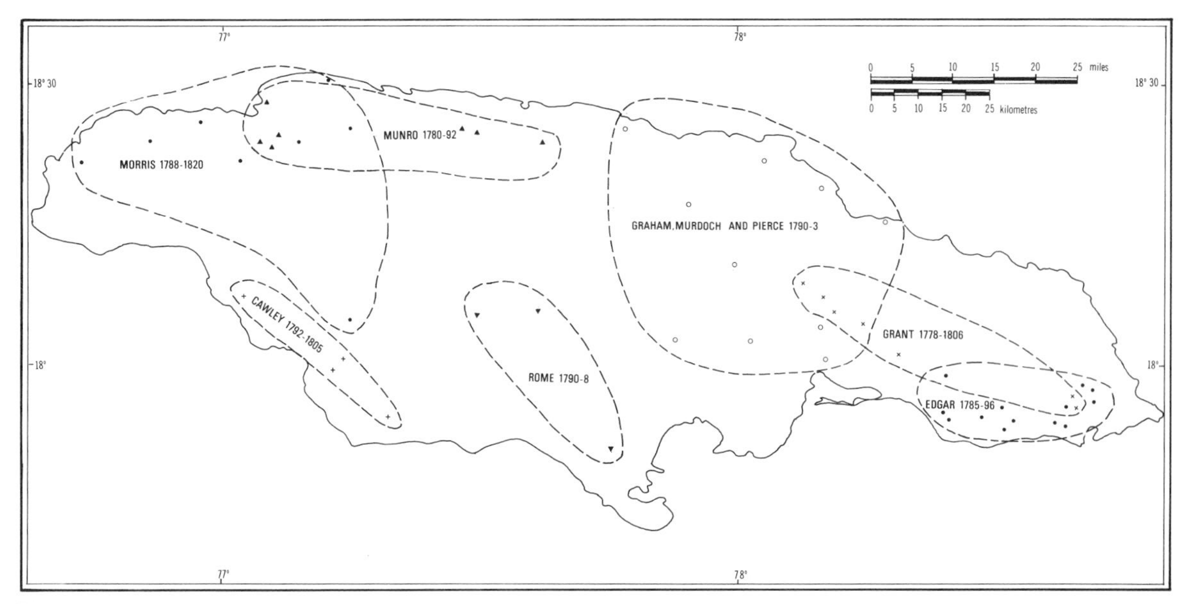

Fig. 5.2 Regional patterns of surveyors' activities about 1790

emergence of regional spheres of activity was similar to that of Great Britain at the same period, but most closely parallel to the situation in Ireland.[24] J. H. Andrews has argued that the estate map was the most "characteristically Irish" product of Ireland's planmakers, and that estate surveyors were the only cartographers permanently resident in Ireland. There, says Andrews, the method of making estate maps "was transmitted inside the country and often inside the family by men who were generally without experience of land surveying as carried out across the Channel." But their "national roots" did not go very deep, it seems, "for one finds no major regional variants (such as might be expected to have developed in the course of a long process of evolution), and English personal names become common as the map-making profession is traced backwards, until a probable line of descent is picked up running through the official admeasurements carried out by imported surveyors at various times before 1703."[25]

Although Jamaica possessed no real equivalent of Ireland's official "plantation cartography," the course of development from transplanted English practice in the seventeenth century, through two centuries of local practice relatively unaffected by metropolitan advances, was very similar. Succession to the surveying profession through birth, apprenticeship, and partnership created a long line of descent in Jamaica, but, as in Ireland, there seems no clear evidence of regional variation in technique or practice. The origins of the Jamaican lines of descent may well be found in the mid-eighteenth century rather than in the seventeenth century, so the chances of development were limited, and the small size of the island prevented evolution in isolation.

Surveying Techniques

The relative isolation of Jamaica's surveyors during the eighteenth and nineteenth centuries contributed to a slow rate of adoption of metropolitan techniques. Thus the basic instruments and techniques employed by the Jamaican land surveyor remained the compass and the chain at least until the middle of the nineteenth century, while the theodolite and triangulation were little used. For the plantation surveyor, however, compass and chain produced results generally acceptable to the employer, and offered a fair level of accuracy in the measurement of area.

No Jamaican land surveyor sat down to write a manual explaining the techniques he used in his work, so it is necessary to reconstruct the method of procedure from the surveyors' field books, protractions, plans, and inventories, together with less systematic materials. Field books exist in considerable quantity in the National Library of Jamaica's collection of plans, but they have not been catalogued. Very often the surveyor did not sign his field book, and it is difficult to relate field books to particular plans. Moreover, the inventories are generally imprecise when describing surveyors' instruments. The sources are sufficient, however, to permit a general description of the techniques used by Jamaican surveyors after 1750.[26]

Surveying with compass and chain involved moving around the perimeter of each parcel of land, measuring the distance between each station with a chain of fixed length, and, at the same time, taking the angles between successive stations by means of a compass. The chains used in Jamaica after 1700 were apparently of the standard introduced by Edmund Gunter in 1620. They were 66 feet in length and divided into 100 links, each link joined to the next by three rings and every tenth link being distinguished by a piece of brass of different shape to facilitate counting.

One Jamaican plan of 1762 by John Rome specified that it had a "scale of Gunter's chains," and another of about 1820 had a "scale of English chains, each 66 feet." Jamaica developed no creole measures, like the Scots chain of 1.12 imperial chains, or the Irish "plantation perch" of 21 feet, and apart from occasional accusations (leading, in the case of James Robertson, as far as libel actions in the high courts of England), that surveyors used "short chains," the Jamaican chain was standardized after 1700.[27] In order to reduce the measurement of lines to correct horizontal measure, the chain could be held perfectly straight in the horizontal by means of a plummet, but where the land was steep this required an excessive number of short chain measurements. The alternative and preferred method was to measure the vertical angles along the line, using a level, and then to read the horizontal distances from a diagram or conversion table.

To measure angles in the horizontal plane, Jamaican surveyors of the eighteenth and nineteenth centuries generally used a standard surveying compass with sighting vanes or perhaps a plane table marked out as a semicircle. By the early nineteenth century the most important angle-measuring instrument was the theodolite, which had the advantage that it could be used for both horizontal and vertical angles. But Jamaican surveyors rarely recorded angles of less than half a degree, suggesting that the theodolite, which allowed measurement in minutes, remained relatively rare in the island until the later nineteenth century. Neither their field books nor their inventories referred explicitly to theodolites, but they were on sale in Kingston by 1850 and perhaps earlier.[28]

Slaves and laborers employed by Jamaican surveyors were generally given only manual tasks to perform. Most of this work involved carrying goods and instruments and clearing bush, but the "chainers" were depended on for accurate work. The slaves owned by surveyors seem not to have been trained in skilled tasks, but worked strictly as domestics and laborers. In this their position matched that of slaves in the United States, except that the smallness of the Jamaican white population by the middle of the eighteenth century meant that slaves were more often used as chainmen.[29]

In 1842 the committee of the Jamaican Assembly enquiring into the affairs of land surveyors put a tantalizing question to Smith and McGeachy, but got no relevant response. They were asked: "As a knowledge of land surveying can be practically acquired in many of the public and private schools of England or France, why should a service of five years of apprenticeship be necessary, particularly as a branch of surveying is taught in those schools which is, if not unknown, not practised in Jamaica?"[30] The "branch of surveying" not followed in Jamaica cannot be identified certainly, but it seems most likely that the assembly meant triangulation. By the early nineteenth century the accurate measurement of land by triangulation, which had begun in France in the seventeenth century and developed through the British Ordnance Survey after 1780, had ceased to be seen as an aspect of military engineering, and had been adopted by most local surveyors.[31] In England, surveyors needed to use chain triangulation to map the complex tenurial patterns of open fields in enclosure-award work, but this demand was less obvious in Jamaica even in the subdivision of plantation lands after emancipation. Similarly in Ireland, triangulation remained a rarity among estate surveyors until the coming of the Ordnance Survey in the 1820s. Andrews's remark that survey by compass and chain was "regarded outside Ireland as an essentially colonial instrument" seems to fit the Jamaican case.[32] One reason for the slow adoption of triangulation in Jamaica was the general absence of people with military experience among the plantation surveyors.

Major Jean Bonnet Pechon, a French coffee planter and *ingénieur du roi*, came to Jamaica following the revolution in St. Domingue, and signed himself "Knight of the Royal and Military Order of Saint Louis, Engineer."[33] In 1807 he was appointed assistant island engineer after completing a number of plantation surveys. One of the earliest of these, his plan of Belvidere Estate in 1800, was described as having been "geometrically performed," and his 1807 plan of Whitney Estate was "geometrically observed and drawn," hinting the two plans may be based on triangulation, though his other plans do not make such claims. These two plans bear the marks of unusual precision, and Pechon's use of hachures betrays his training in military survey techniques (figure 5.3), but it is impossible to demonstrate that these plans were in fact based on triangulation, since the associated field books and protraction sheets seem not to have survived.

Apart from these plans by Pechon, there is little evidence to suggest that triangulation was employed by Jamaica's plantation surveyors, and few of them had any acquaintance with military procedures. The influence on Thomas Harrison of his work in Panama under U.S. engineers is not obvious in his plantation cartography, though it no doubt contributed to his late surveys for Jamaican railroad and canal routes. In general it was rare for engineers to be employed in plantation surveying in Jamaica, and unusual for local surveyors to perform topographic work. A similar division occurred within the Irish surveying profession in the early

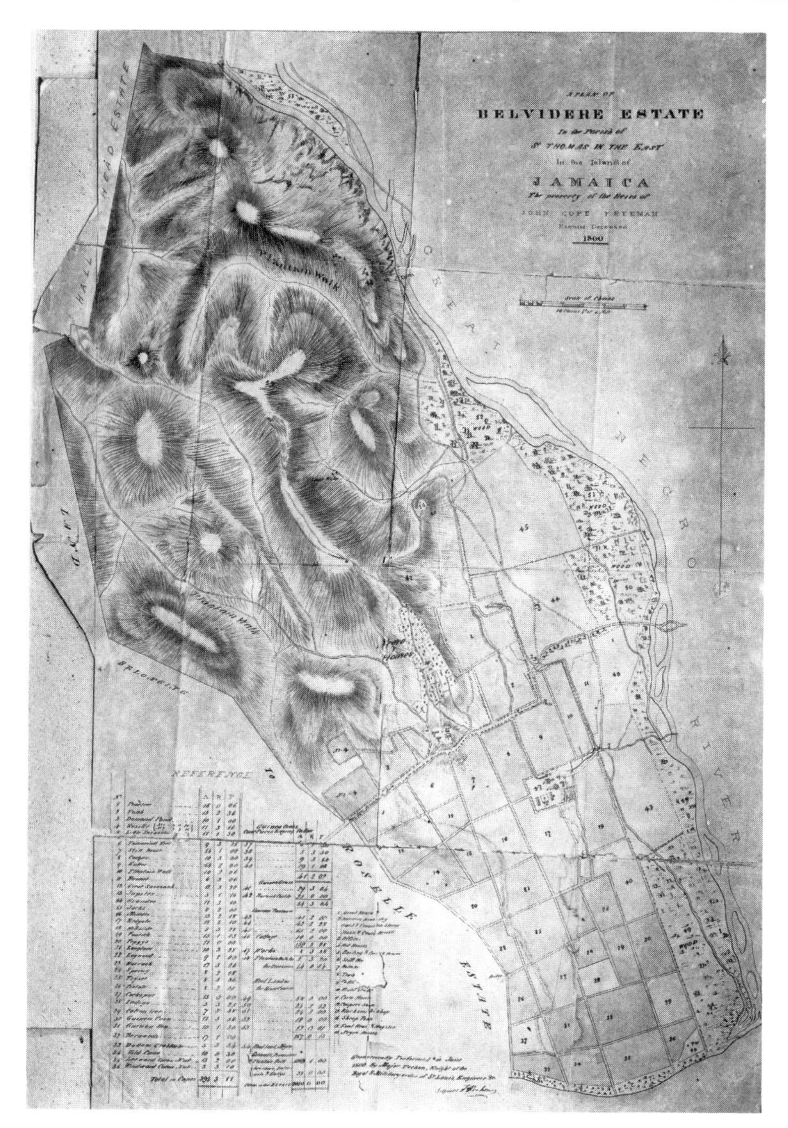

Fig. 5.3 Jean Bonnet Pechon, plan of Belvidere Estate, 1800 (Courtesy of the National Library of Jamaica)

nineteenth century because, Andrews suggests, "the engineer was not cheap enough for estate surveying, and the estate surveyor not skilled enough for trigonometry or topography." In colonial America there appears to have been a greater overlap between estate surveyor and regional cartographer.[34]

In Jamaica, surveying and planmaking procedures were clearly separated. Often the fieldwork was performed by the senior partner of a surveying firm, while the translation of the booked measurements into a final plan was left to a junior assistant or apprentice. The copying and reduction of plans was almost always the work of assistants; further, the production of plantation maps did not always depend on measurements taken by "actual survey," but combined information from a variety of sources, such as registered plats and papers surviving from former surveys. This was particularly the case when planmaker wished to map larger regions and became a regional cartographer.[35]

The National Library of Jamaica Plan Collection

The collection of plans in the National Library of Jamaica originated in the gradual accumulation of surveyors' papers, and so contains a large proportion of unfinished plans, the more polished pieces of work having generally found their way into the hands of the plantocracy. The presence of these working papers is a great advantage for the study of surveying and planmaking techniques, but it also means that many of the plans are too unfinished to be fully interpretable.[36]

The collection contains about twenty thousand items, and of this total nearly one thousand portray the internal land-use pattern of plantations. Some are copies and duplicates, however, leaving us with 856 unique plantation maps; this is a substantial sample, adequate for many types of analysis. In addition, an unknown number of estate maps exist outside the National Library, many of them the work of surveyors and planmakers identified here and in chapter 6. No attempt has been made systematically to incorporate these plans in the analysis.

It is difficult to assess the representativeness of the plantation maps in the National Library. Certainly there is no plan portraying internal land-use patterns for every estate, plantation, and livestock pen, though simple plats are available for a large proportion of them. Some planters and penkeepers, particularly the resident owners of relatively small holdings, never bothered to make detailed plans of their properties. Other plantations, however, were surveyed frequently. The total number of estates, plantations, and pens operating in Jamaica varied over time, but a useful point of comparison may be found in the number functioning around 1800.

Robertson's map of 1804 included 826 sugar estates, while the total number of plans for "estates" during the period 1700–1839 amounts to 319. This suggests that roughly 40 percent of Jamaica's sugar estates are represented for this period. For the later nineteenth century, the earliest count of sugar estates relates to 1881, and this suggests that as many as 61 percent were represented by plans in the period 1839–1900, though the rapid decline in the sugar economy after 1838 means that this estimate is less valuable. In the period of slavery, plans exist for roughly 20 percent of the plantations producing coffee, pimento, and other staples, and for 30 percent of the pens. The sugar estates are therefore better represented than the plantations and pens, reflecting their social and economic dominance.[37]

Only 21 of the 856 estate maps are dated earlier than 1779, and only 18 relate to the period after 1870; only 2 are earlier than 1750. Thus more than 95 percent of the plans are from the period 1770–1870. Figure 5.4 charts the number of plans produced in each year for this period of major activity, but excludes plans that could be dated only to the nearest decade (thus removing misleading peaks). Within this century of surveying activity, several periods are clearly defined. The first, from 1770 to 1790, showed a rapid growth in the number of plans portraying the internal layout of plantations. This growth was in part a product of the development and spread of the sugar industry, but it also reflects

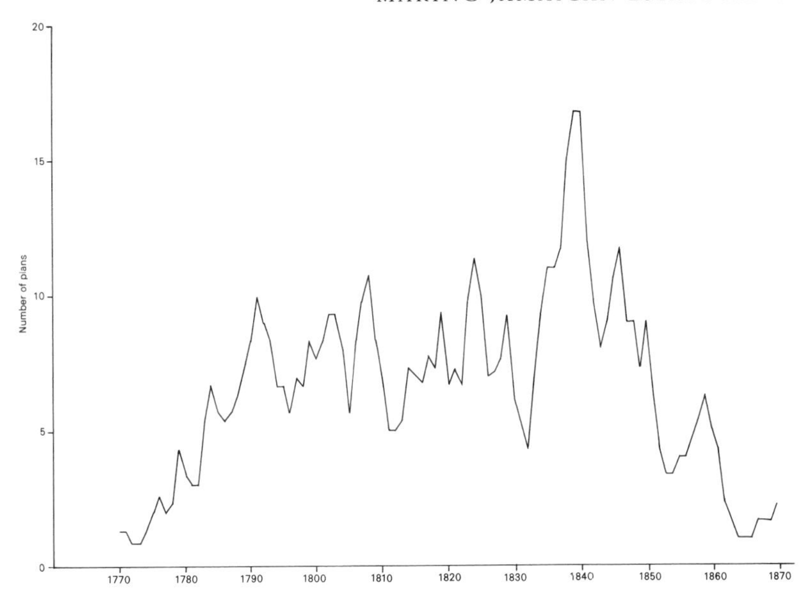

Fig. 5.4 Estate maps produced yearly in Jamaica, 1770–1870

developments in surveying that made the mapping of internal rather than mere property boundaries increasingly common.

The second period, 1790–1834, was marked by a steady output of plans, decline in the sugar industry being balanced by growth in coffee and the demand for surveys of newly established plantations. In the third period, 1834–50, surveying activity reached a high pitch in the plantation sector, with a very pronounced peak in the years 1838 to 1840. The coming of emancipation brought a desire by planters to obtain accurate surveys of their properties, to meet the demands of a wage labor system, and to assess the extent and value of lands that the slaves had been allowed to cultivate under only limited supervision, but that came to form potential rentable land with the abolition of slavery.

During the 1840s the abandonment and subdivision of plantations meant that the demand for surveyors was maintained at a high level. The fourth and final period, 1850–70, was characterized by steady decline, as the contraction of the plantation system continued and the number of estates diminished rapidly. After 1870 these trends persisted, and at the same time Harrison's cadastral maps did something to reduce the demand for plantation surveys.

The identification of these trends is based solely on those plans that happened to survive. But it is improbable that the actual pattern was greatly different, since the destruction of plans was not determined by age, and surveyors generally recognized that "old plans" might be just as valuable in their work as the most recent. Certainly the survival of plans seems not to increase with proximity to 1900; indeed, the paucity of post-1870 plans may be explained in part by a declining need to collect plans from private hands, as the cadastral survey and the public Survey Department flourished.

The period of greatest plantation surveying activity in Jamaica, 1780–1850, matched closely what Harley has called "the golden age of the local land surveyor" and Thompson "the golden age of the chain surveyor" in Great Britain and Ireland.[38] The principal reasons for this development in Britain were the great demand for the services of professional surveyors created by enclosure, clearance, and general estate "improvement," as well as industrialization and urbanization, in the century before 1850 when the Ordnance Survey killed chain surveying.[39] In Jamaica, changes in plantation management techniques together with the challenge of emancipation created an equivalent need for plantation surveys, and Harrison's cadastral maps played a role in diminishing the demand for property surveys not unlike that of the Ordnance Survey. Jamaica's surveyors, unlike those of Britain and Ireland, were rarely employed as estate managers, probably because it was unusual for planters to depend on rents collected from tenants, and so they had only a limited requirement for land-valuation services.[40]

Cartographic Techniques

Almost all of the plans were on paper, and very few of those preserved in the National Library formed part of atlases; there were few Jamaican counterparts to the fine atlases of estate lands in bound volumes, common for eighteenth-century Scotland and Ireland.[41] The dimensions of the plans varied considerably but generally decreased over the period from 1750 to 1890; thus the typical plan of 1770 was roughly three times the size of that of 1870.

The outlines of most Jamaican plantation maps were rendered in Indian ink and embellished with watercolors. A small proportion were in pencil, most of these being rough drafts, and some plans contained elements of each of these media. Many early planmakers used color only to highlight the outlines of features within plantations. The plan of Parnassus Estate based on James Cradock's survey of 1758 is typical of this style, and of all the plans in the Dawkins Plantation papers from which it comes (plate 5). The overall impression given by this plan is a subtle celadon green, but the roads are colored light amber and the works buildings red. Although the roads are blocked in, the cane fields are left empty except for their borders, and only the waste land on the river bend is fully colored green, while the pastures are distinguished by symbolic copses. Thomas Harrison's 1850 survey and plan of Norris Estate, for McGeachy and Griffiths, combines use of a wider range of colors with a great deal of work in pen and ink (figure 5.5). This plan is in fact a highly developed protraction, the meridian lines in red ink and the stations of survey indicated by circles and dots. Only the cane fields and pastures are fully colored, and it appears that the gradations of green used for the former were intended to distinguish adjacent field units rather than specifying anything about their contents. In general, color was used in a highly pragmatic and individualistic fashion, but it was made to serve the function of conveying information rather than merely being used as a decorative aid.

Planmakers chose scales for their maps according to the size of the plantations and the amount of detail they wished to include; there was no significant correlation between scale and the total area of the sheet used. Many Jamaican planmakers failed to include any indication of the scale employed in their plantation maps. For some plans the scale is known only because it was added by a later surveyor, most commonly during the preparation of the parish cadastral maps at the end of the nineteenth century. In all, only 59 percent of the 856 plans studied carry scales, but this percentage

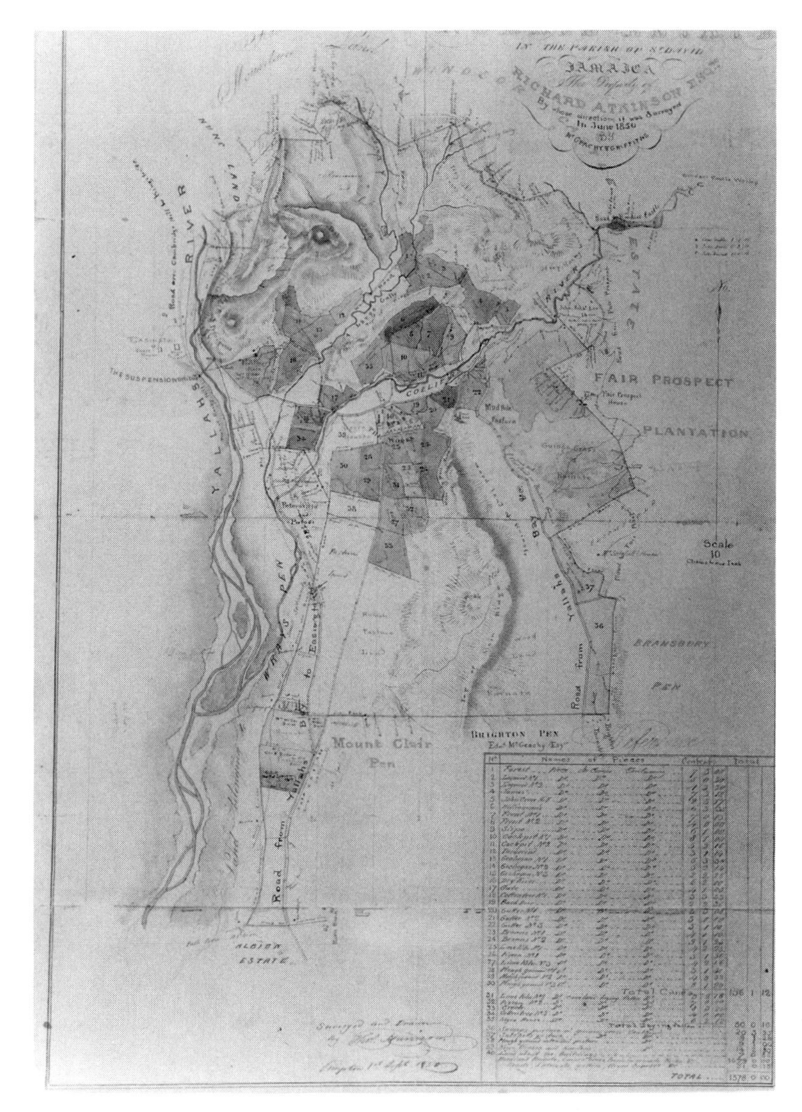

Fig. 5.5 Thomas Harrison, Norris Estate, 1850 (Courtesy of the National Library of Jamaica)

tended to increase toward the time of emancipation, suggesting a gradual refinement of technique.[42]

The great majority of Jamaican estate maps were drawn with north at the top of the sheet, the meridians parallel to the sides of the paper. The "top" of the sheet can be identified by the general direction of the writing on the plan, and only in rare cases does this definition create any confusion. Some planmakers oriented their plans with west at the top in order to satisfy a preference for particular dimensions, and a few used meridian lines that were not parallel to the edge of the sheet; few distinguished between magnetic and true azimuth.[43]

Reference tables listing the names, numbers, and areas of fields and other features are more common than scales and north points. As many as 86 percent of the plans are equipped with such tables, suggesting the central role that they played. This proportion varies little from period to period, never falling below 75 percent. Most of the plans lacking reference tables are incomplete rough drafts or protractions, but in some cases most of the information normally found in marginal tables is located on the plan itself.

The most common form of table comprises a list of numbers or letters, related to field names and their areas, with only the numbers or letters appearing on the plan; some 58 percent of the reference tables are of this type. This style persisted through out the eighteenth and nineteenth centuries, and is well illustrated by the 1758 plan of Parnassus Estate (plate 5). On this plan the cane fields are designated by capital letters, named (with two exceptions) in the reference table and termed "pieces." Small fields under pasture around the works, gardens, and stockyards are labelled with lowercase letters and not named in the reference table. The numbers 1 to 8 designate large fields under pasture, and 9 to 16 provision grounds, grass, waste land, the slave houses, and "land in the possession of free negroes," none of these units being named. The acreage of each unit is also shown in the table, and subtotalled for each type of land use. The practice of naming only fields planted in the main crop also applied on coffee plantations, while pastures were usually named only on pens. The main modification to the system employed on the Parnassus plan was the use of continuous numbering, which became increasingly common after about 1830.

Reference tables are generally simple and austere. Some consist merely of lists, but most enclose the columns within a frame, partitioned to distinguish field numbers, names, and units of area. The table on the Parnassus plan is typical of the latter style. In the eighteenth century, however, a good number of planmakers enclosed their tables within elaborate cartouches, borrowing their iconography from contemporary architecture, furniture, and memorial tablets, as well as from traditional regional cartography.

Although most plantation maps provide at least rudimentary representation of watercourses, either as single or parallel lines, other features of the landscape are generally neglected. Some maps depict morasses and swamps where these were extensive. The plan of Parnassus is typical of this style. Only 22 percent of the plans contain any indication of relief; before 1800 this proportion is as small as 13 percent, and even after 1850 it rises only as high as 47 percent.

Jamaican surveyors were poorly equipped to measure relief before about 1850, topography generally being the province of engineers and regional cartographers. Detailed levelling work and the plotting of sections are performed only for constructing roads, railways, and canals. Relief was also a secondary concern for many planters. Most sugar estates were located on relatively level plains, and even where they incorporated hilly terrain, the cane fields themselves were largely confined to pockets of flat or gently sloping land. Coffee plantations, however, were usually found in mountainous areas, and the crop was planted on steep slopes. Thus it is not surprising that indications of relief appear more often on plans of coffee plantations than on plans of sugar estates. Over the whole period 28 percent of plans of coffee plantations contain some form of relief representation, compared to only 22 percent for sugar estates, and 14 percent for livestock pens. For a smaller sample of 282 relatively complete plans it can be shown that there is also a direct relationship between altitude and the representation of relief. The critical point occurred at around 2,000 feet above sea level; plans located below this elevation represent relief in only 19 percent of cases, while 34 percent of plans for plantations above 2,000 feet depict relief in some fashion.

Of the plans that do depict relief, 68 percent use only hachures, 23 percent hill shading with some hachuring, and 7 percent perspective drawing. The last of these methods was popular during the eighteenth century, but little used after the abolition of the slave trade.[44] In most cases, as in the plan of Mount Ephraim, the perspective is from the bottom of the sheet, and some pretense was made to draw the hills in a realistic pattern (figure 5.6). Although the technique of perspective drawing is rare after 1800, it does occasionally appear as an archaism.

Hachures, applied either by pen or by brush, often provided the only indication of relief on Jamaican plantation maps. An example of simple hachuring, of the type often called "hairy caterpillars," is provided by Griffiths's 1860 plan of Lindale, while the protraction of the survey on which this plan was based demonstrates the crudity of the raw material (figures 5.7 and 5.8). The engineer Pechon's 1800 plan of Belvidere shows a more polished use of the technique, though the precision of the representation is probably no greater (figure 5.3). More obviously exact is a plan of Windsor Pen, Trelawny, made in 1795 by John Henry Schroeter (figure 5.9). Windsor was embedded in the margins of the Cockpit Country, and Schroeter's fine black

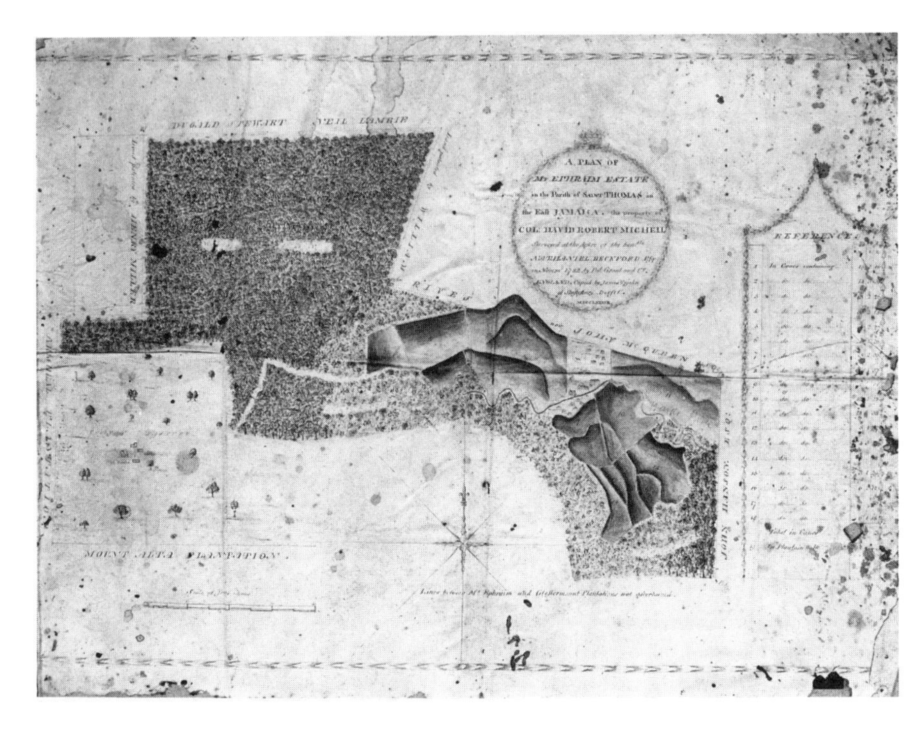

Fig. 5.6 Patrick Grant, plan of Mount Ephraim Estate, 1782 (Courtesy of the National Library of Jamaica)

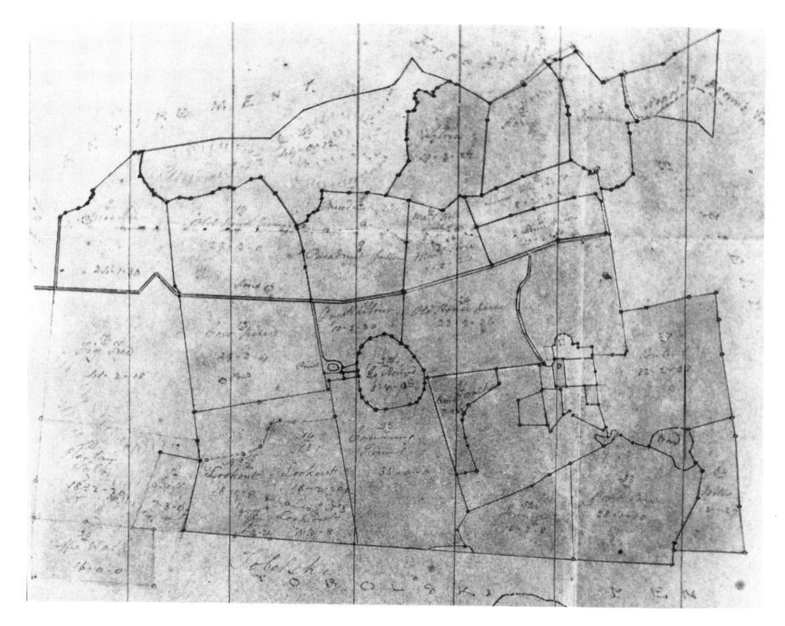

Fig. 5.7 M. J. Griffiths, protraction of survey of Lindale Pen, 1860 (Courtesy of the National Library of Jamaica).

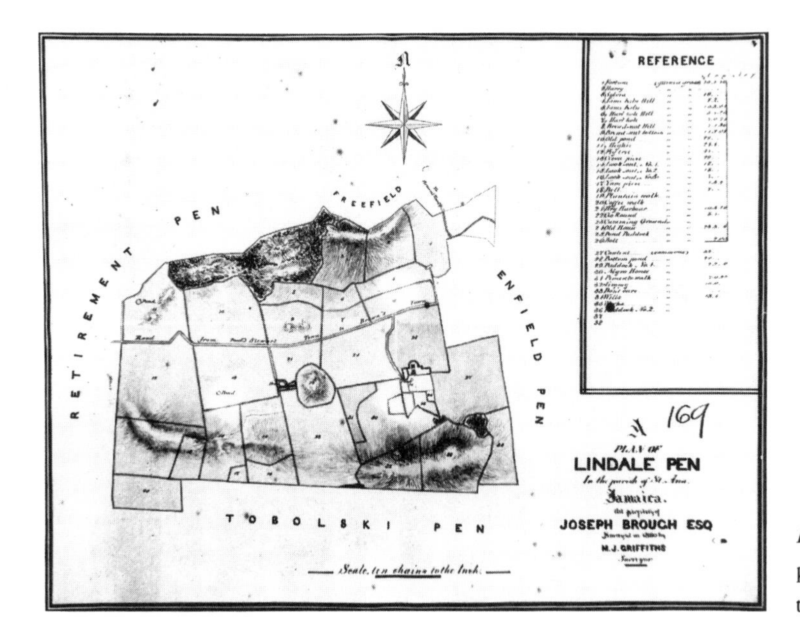

Fig. 5.8 M. J. Griffiths, fair copy of plan of Lindale Pen, 1860 (Courtesy of the National Library of Jamaica)

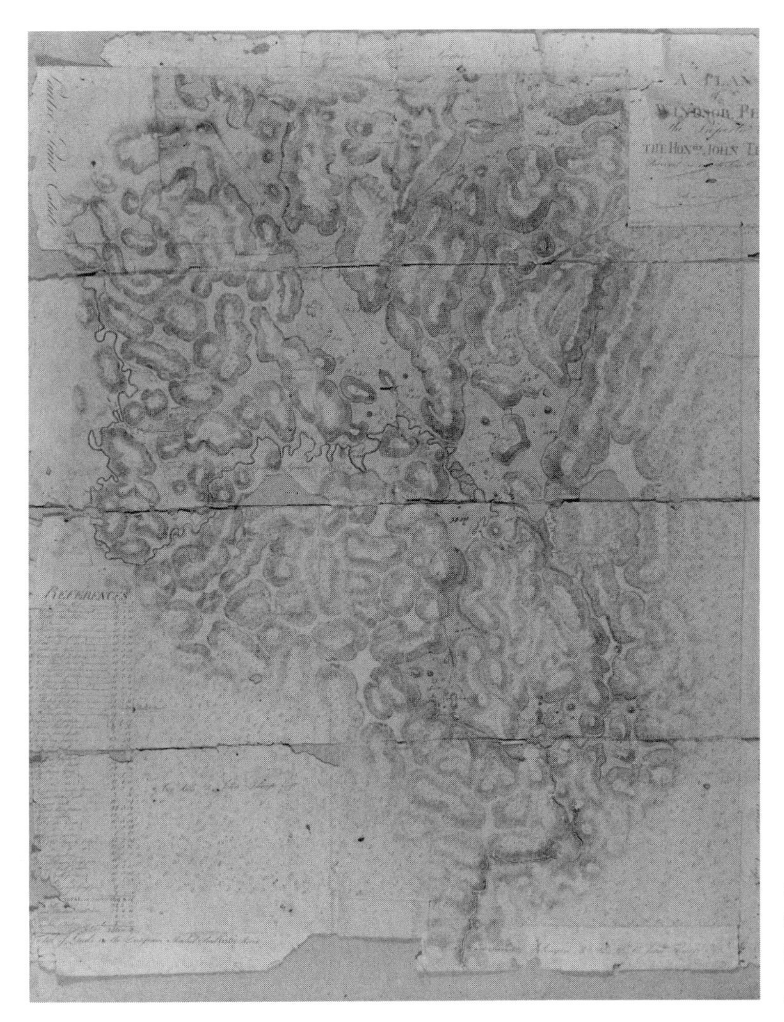

Fig. 5.9 John Henry Schroeter, detail from a plan of Windsor Pen, 1795 (Courtesy of the National Library of Jamaica)

hachures bring out the characteristics of the karst landscape in striking fashion. Schroeter served as captain at Fort Balcarres in nearby Falmouth, and his military training is also evident in his earlier plan of a road in St. Ann, one of the few plans to show the true azimuth as well as the magnetic meridian.[45] The precision of his plan of Windsor seems almost to have appalled Thomas Harrison, who wrote on the reverse, "A remarkable specimen of patience."[46]

Only one Jamaican plantation map produced before 1900 shows the beginnings of an attempt to employ contour lines (which entered European cartography in the late eighteenth century), and these lines are not given any quantitative value. This is Thomas Harrison's 1877 plan of Rozelle in St.

Thomas, a sugar estate with level cane fields on its seaside boundary and hills rising above 1,000 feet in its northern backlands (figure 5.10). Harrison placed contour lines on the plan but used hachures to indicate the steepest slopes. At the end of the reference table he noted, "The topography obtained from numerous sketches on the spot." Comparison with the modern topographical sheet suggests a fair degree of accuracy, but the contours are drawn too softly to provide a striking impression of relief.

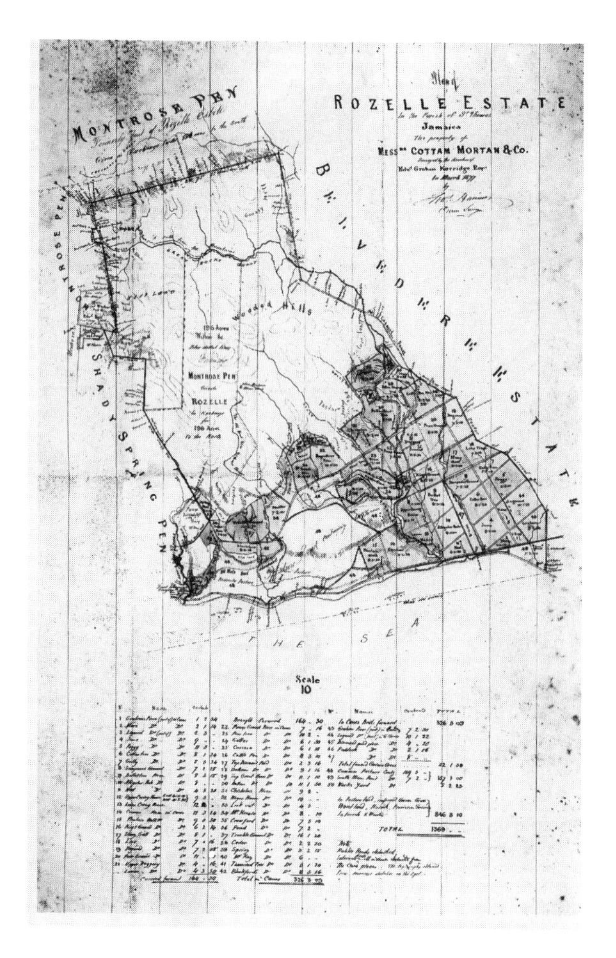

Fig. 5.10 Thomas Harrison, plan of Rozelle Estate, 1877 (Courtesy of the National Library of Jamaica)

Jamaican planmakers employed a wide range of symbols to represent features of plantation land use, rarely achieving a high degree of standardization. Some elements of the symbolic systems used have been discussed already; thus the use of color to represent different types of land use has been shown to vary from surveyor to surveyor and plan to plan, though most made some effort to respect naturalistic associations. Most often, however, fields were left blank and their land use identified through reference to marginal tables. Only for 35 percent of the plans is land use symbolized by color, texture, or pictograph.

The most important development in methods of symbolization during the eighteenth and nineteenth centuries was the replacement of perspective drawing with features shown in plan.[47] This development has already been considered in the context of relief representation, but similar principles applied more generally. In part, the gradual abandonment of pictorial symbols was a product of the decreasing reliance on the perspective of the landscape painter. Thus the technique of representation employed serves as a guide to the method of survey.

The representation of plantation buildings provides a clear example of this development. Many Jamaican plantation maps of the eighteenth century use perspective drawing for all buildings, but without attempting to maintain a consistent perspective, merely displaying the facade of each building. The next stage, beginning in the eighteenth century, combines perspective drawings of great houses and slave villages with planform depiction of works buildings. In the final stage all plantation buildings are shown in plan. The 1787 plan of Mount Ephraim illustrates the first stage, all buildings appearing as perspective drawings, just as relief is presented on these maps (figure 5.6). The second technique appeared as early as 1758 in Cradock's plan of Parnassus (color plate 5) and persisted at least until the 1840s. The consistent representation of all buildings in plan was not unknown in the eighteenth century, but became increasingly common after 1800. This somewhat erratic progression toward precision suggests that surveyors were first required to plot works buildings accurately, and only later began to map the houses of planters and workers in plan, either because they had in fact measured them, or because they wished to produce a coherent plan.

Precise locating of trees was even less important to surveyors and planters than the layout of workers' houses. Thus trees continued to be represented by pictographs throughout the entire period, and other elements of the landscape not actually measured during survey were treated in the same way. The only really significant change in technique was that, while some early plans failed to adopt a consistent perspective, those of the nineteenth century were all viewed from a single point, the bottom of the sheet, with shadows cast by a sun to the left, usually the west. There was also a tendency for trees to be homogenized and ordered into geometric patterns, so that they gradually came to resemble symbolic shading systems or even mechanical stipples, rather than random individual pictographs. Only the occasional map represents such patterns in plan; as in the representation of buildings, this development did not fit a chronological pattern, archaic and precocious techniques overlapping a good deal.

Cultivated crops are generally represented by colors or blank spaces, while pastures and woodland are distinguished by symbolic copses, isolated trees, and swards. Coffee fields are covered in symbolic coffee trees on a few plans, and on others coffee trees are interspersed among shade trees, whereas the plantations' woodland is left blank.[48] On Francis Ramsay's 1810 plan of Clifton Mount Plantation, the coffee trees are arranged so as to represent a stipple, while a variety of food crops can be identified in the provision grounds by means of the naturalistic drawings. Fields of cane and grass are also represented by geometric or curvilinear patterns on some plans. Guinea grass is distinguished by parallel ruled lines on the 1800 Cedar Valley plan, and by a less formalized pattern of lines on the 1780 Goshen plan. Frazer and Rainey's plan of White Hall Estate dated 1788 is unusual in its use of stripes and intervening dotted lines to symbolize trenched cane fields, creating patterns reminiscent of ridge-and-furrow open fields. These patterns, together with the cultivated strips representing provision crops scattered among the estate's woodland, were not intended to suggest more than generalized indications of cultivation practices, and must be understood in terms of the artistic elements of cartographic style as a whole.

Changes in the lettering of plantation maps reflect broader tendencies in calligraphy, typography, and the lettering of monumental inscriptions.[49] Eighteenth-century planmakers sometimes lettered their titles and marginal tables in cursive script, using a quill and imitating the more elaborate efforts of the writing masters. Others combined calligraphy with letters taken from copybook alphabets of different and not always compatible styles.

The use of alphabets became increasingly common after 1800. When Patrick Keeffe died in 1822, the surveyor had among his tools of trade four patent alphabets in books, one marked "No. 1" and another "No. 7," worth almost £10.[50] Later in the century, Jamaican planmakers introduced to their titles ornamental alphabets typical of Victorian advertising typography. The names of bounding plantations were often lettered in such distinctive styles, though most of the detailed lettering of the internal layout of the plantation and its reference table continued to be done freehand.

The cartouche used to enclose the title or reference table of a map had its origins in the oval or oblong figures in Egyptian hieroglyphics that surround inscriptions. By the eighteenth century the cartouche usually consisted of an oval tablet with ornamental scrollwork, an oval formed by a rope knotted at one end, or a tablet representing a sheet of paper with the ends rolled up.[51] Cradock's plan of Parnassus in 1758 provides an archetypal example of the oval tablet enclosed within scrollwork (plate 5), and this style was carried into the early nineteenth century. By the 1780s planmakers began to simplify the oval tablet, providing a variety of other forms of embellishment on their plans. In the nineteenth century, when tablets and embellishments were completely displaced by lettering in the titles of Jamaican plantation maps, particularly after about 1830, some planmakers retained echoes of the style by forming the name of the plantation or its owner into a convex or concave crescent. In some eighteenth-century plans the oval tablet disappeared while the scrollwork remained, sometimes assuming the alternative cartouche shape of a sheet of paper rolled at the ends. This last style occurs on Jamaican plantation maps from about 1780 to 1830, becoming increasingly solid and reminiscent of monumental masonry rather than paper. Other planmakers of the late eighteenth century favored festoons and cloth drapes (figure 5.6). Mythical animals and alligators derived from the Jamaican coat of arms top some cartouches from the same period, while a few nineteenth-century examples include the coat of arms of the planter-proprietor.

Some cartouches are overwhelmingly pictorial, but these rarely provide useful historical data since they are essentially symbolic and decorative in content. Only in a few cases do they contain representations of people performing work, or of houses and tools, for example; more often they are dominated by leafy vegetation. Valuable pictorial material is in the margins of some plantation maps. The most useful is that which depicts plantation buildings, providing elevations to compensate for the loss of information that resulted from the shift from perspective drawing to planform mapping. The bucolic scenes that decorate the corners of some plans are much less useful, since their locations cannot be identified and they appear to have more to do with Arcadia than with the slave society of Jamaica.

Conclusion

The surveying and planmaking techniques employed by Jamaica's land surveyors were not particularly backward by comparison with their metropolitan fellows. The great demand for precise survey created by enclosure, agricultural improvement, and industrialization in England had no real parallel in Jamaica. In the eighteenth and nineteenth centuries England's local surveyors generally reverted to surveying by the chain alone because of the high cost of the new improved angle-measuring instruments required for triangulation, such as theodolites, and because of their limited education. Only in the later nineteenth century did instrument construction and scale division reach a high degree of perfection, and the standardization and simplification that were necessary to make the instruments cheap enough for the average land surveyors took even longer.[52]

Thus Thompson has rightly termed the period 1750–1850 "the golden age of the chain surveyor" in Britain and Ireland.[53] Although Jamaican surveying was backward in the sense that it lacked an ordnance survey and failed to practice triangulation, the local plantation surveyor employed methods that were hardly inferior to those of his contemporary metropolitan estate surveyors, in spite of his colonial isolation. Nor were Jamaican surveying techniques inferior to U.S. practice, where the theodolite was little used in the nineteenth century and chain and compass reigned supreme.[54] Further, although the techniques of the plantation surveyor were inadequate for precise cadastral and engineering work, they did yield good results for plantation purposes, especially after 1750, and the value of the surveyors' estate maps for understanding land-use patterns was certainly sufficient to ensure the patronage of the plantocracy.

Notes

A few sentences are taken verbatim from *Jamaica Surveyed* and the *Journal of Historical Geography* 13:17–39, but the text has been fully rewritten and reorganized and includes much new material.

1. V. B. Grant, *Jamaican Land Law* (Kingston, 1957), 1–7; Lloyd G. Barnett, *The Constitutional Law of Jamaica* (Oxford, 1977), 1–12.

2. *Jamaica House of Assembly Journals* 1 (6 September 1683):68; (8 September 1683):70, Act 24 (1683).

3. Act 14 (1785) and 31 Geo. 3, c. 22; *Jamaica House of Assembly Journals* 14 (27 November–5 December 1823):201–14 notices an unspecified amendment.

4. *Jamaica House of Assembly Votes*, 1842, 232–47 (Appendix 23), and 22 Vic., c. 40. The 1858 act stated "that the several enactments in relation to surveyors of land should be repealed" but failed to specify those then in force.

5. Law 13 (1869).

6. See E. G. R. Taylor, *The Mathematical Practitioners of Tudor and Stuart England* (Cambridge, 1954); Taylor, *The Mathematical Practitioners of Hanoverian England* (Cambridge, 1966); F. M. L. Thompson, *Chartered Surveyors: The Growth of a Profession* (London, 1968).

7. It is assumed that the average annual number of commissioned surveyors was ten for 1700–1780 and twelve for 1847–81, and that that average working life of each surveyor was ten years.

8. Ian H. Adams, "The Agents of Agricultural Change," in *The Making of the Scottish Countryside*, ed. M. L. Parry and T. R. Slater (London, 1980), 167; Ian H. Adams, "The Land Surveyor and His Influence on the Scottish Rural Landscape," *Scottish Geographical Magazine* 84 (1968):248.

9. See Thompson, *Chartered Surveyors*, 269–73.

10. *Jamaica Almanack*, 1811, list of public officers.

11. Ibid., 1838, 56; *Jamaica House of Assembly Votes*, 1842, 242 (Appendix 23).

12. *Handbook of Jamaica*, 1883, 110; *Daily Gleaner* (Kingston), 26 October 1894; Vincent John Marsala, *Sir John Peter Grant; Governor of Jamaica, 1866–1874* (Kingston, 1972), 105.

13. *Departmental Reports, Jamaica*, 1892/93, 92; 1893/94, 204; *Handbook of Jamaica*, 1917, 642–43; *Daily Gleaner*, 9–11 October 1916.

14. Censuses of 1844, 1861, and 1891.

15. *Jamaica Critic* 4 (May 1929):49–50; *West Indian Critic and Review* 6 (April 1921):43; 6 (1931):4–6.

16. *Jamaica Almanack*, 1799, 95; 1838, 43.

17. 22 Vic. c. 40, Law 31 (1894).

18. The inventories are fully analyzed in B. W. Higman, *Jamaica Surveyed: Plantation Maps and Plans of the Eighteenth and Nineteenth Centuries* (Kingston, 1988), 31–38.

19. B. W. Higman, *Slave Population and Economy in Jamaica, 1807–1834* (Cambridge, 1976), 69.

20. Higman, *Jamaica Surveyed*, 31–38.

21. For a fuller discussion of these features, see ibid., 39–43.

22. See table in ibid., 41.

23. See table in ibid., 42.

24. F. W. Steer, "A Dictionary of Land Surveyors in Britain," *Cartographic Journal* 4 (1967):124–26.

25. J. H. Andrews, "The French School of Dublin Land Surveyors," *Irish Geography* 5 (1967):277. See also Andrews, *Plantation Acres: An Historical Study of the Irish Land Surveyor and His Maps* (Belfast, 1985).

26. See Higman, *Jamaica Surveyed,* 49–59, on surveying techniques.

27. Plan of Caymanas, 1762, National Library of Jamaica, Dawkins MS, vol. 6; plan of Springfield Estate, c. 1830, National Library of Jamaica plans, Hanover 1; Edmund Gunter, *Use of the Sector, Crosse-Staffe, and Other Instruments* (London, 1624; reprint, Amsterdam, 1971); Betty M. W. Third, "The Significance of Scottish Estate Plans and Associated Documents," *Scottish Studies* 1 (1957):39–64; *Robertson and Stephenson: Appellants and Respondents' Cases, an Appeal to the King in Council from Jamaica* (London, 1805).

28. Higman, *Jamaica Surveyed,* 51.

29. See Walter W. Ristow, ed., *A la Carte: Selected Papers on Maps and Atlases* (Washington, 1972), 115.

30. *Jamaica House of Assembly Votes,* 1842, 235, 237.

31. W. A. Seymour, ed., *A History of the Ordnance Survey* (Folkestone, 1980), 9; Lloyd A. Brown, *The Story of Maps* (New York, 1947), 246–65.

32. J. H. Andrews, *A Paper Landscape: The Ordnance Survey in Nineteenth-Century Ireland* (Oxford, 1975), 10; Joseph W. Ernst, *With Compass and Chain: Federal Land Surveyors in the Old Northwest, 1785–1816* (New York, 1979); Sarah S. Hughes, *Surveyors and Statesmen: Land Measuring in Colonial Virginia* (Richmond, 1979).

33. Philip Wright, ed., *Lady Nugent's Journal* (Kingston, 1966), 311; David Patrick Geggus, *Slavery, War, and Revolution: The British Occupation of Saint Domingue, 1793–1798* (Oxford, 1982), 219.

34. Andrews, *Paper Landscape,* 11; William P. Cumming, *British Maps of Colonial America* (Chicago, 1974), 20.

35. On the actual process of mapmaking, see Higman, *Jamaica Surveyed,* 59–64.

36. For a fuller description of the collection, see ibid., 64–67.

37. For a map showing the geographical distribution of plans of plantations, see ibid., 65.

38. J. B. Harley, *Maps for the Local Historian: A Guide to the British Sources* (London, 1972), 24; Thompson, *Chartered Surveyors,* 32.

39. Thompson, *Chartered Surveyors,* 33–39; Andrews, *Paper Landscape,* 10; Third, "Significance of Scottish Estate Plans," 39; Ian H. Adams, *The Mapping of a Scottish Estate* (Edinburgh, 1971), 1–3.

40. Thompson, *Chartered Surveyors,* 25–32; W. A. Maguire, *The Downshire Estate in Ireland, 1801–1845: The Management of Irish Landed Estates in the Early Nineteenth Century* (Oxford, 1972) 197; T. W. Beastall, *A North Country Estate; The Lumleys and Saundersons as Landowners, 1600–1900* (London, 1975), 213.

41. See Higman, *Jamaica Surveyed,* 68–78.

42. For further details on scale see ibid., 69.

43. The true azimuth appears on J. H. Schroeter's plan of the Carriage Road from Retreat to Chippenham Park, c. 1780 (National Library of Jamaica plans, St. Ann 21).

44. Cf. P. D. A. Harvey, *The History of Topographical Maps: Symbols, Pictures, and Surveys* (London, 1980), 182.

45. See note 43 above. Cf. R. A. Skelton, "The Military Survey of Scotland, 1747–1755," *Scottish Geographical Magazine* 83 (1967):5–16.

46. On the technique of layering, see Higman, *Jamaica Surveyed,* 72–73.

47. Cf. R. A. Skelton, *Maps: A Historical Survey of Their Study and Collecting* (Chicago, 1975), 20.

48. Plan of Wakefield Plantation, 1837, National Library of Jamaica plans, St. Thomas 1C.

49. Cf. Alan Bartram, *Tombstone Lettering in the British Isles* (London, 1978).

50. Inventories, Liber 137, fol. 100 Jamaica Archives, Spanish Town.

51. Brown, *Story of Maps,* 174; Edward Lynam, *The Mapmaker's Art: Essays on the History of Maps* (London, 1953), 54.

52. A. W. Richeson, *English Land Measuring to 1800: Instruments and Practices* (Cambridge, Mass., 1966), 145–61 , 187–88.

53. Thompson, *Chartered Surveyors,* 32.

54. Abel Flint, *A System of Geometry and Trigonometry: Together with a Treatise on Surveying* (Hartford, Conn., 1813), 76.

Chapter SIX

THE EVALUATION *and* INTERPRETATION *of* JAMAICAN ESTATE MAPS *of the* EIGHTEENTH *and* NINETEENTH CENTURIES

B. W. HIGMAN

Maps as Social and Economic Indicators

The study of any map can be approached from two distinct directions. The map is both medium and message. On the one hand, it is an *object* worthy of study in its own right, in terms of cartographic technique and style, to be set within the context of the development of graphic and land-surveying arts. On the other, it is an *information source* providing data on the spatial relationships of selected elements of real life. Each of these perspectives is important, and the historian must take special care to consider the ways in which one affected the other. The planmaker's purpose and cartographic style certainly influenced the nature of the data included in maps of plantations, and it is important to assess this influence in attempting to extract historical information from them. Similarly, the physical character of the plantation in its ecological setting influenced the structure and style of the plan. It is important to be aware of these interactions and their potentially distorting effects at all points.[1] Chapter 5 was concerned with the Jamaican estate plan as an object and the light it can throw on the history of colonial surveying and cartographic techniques. In this chapter I will focus on the estate plan as an information source and the data it can offer concerning plantation economy and society, with special attention to the advantages and disadvantages of the plan as historical evidence.

Insofar as the behavior of Jamaican planters and workers had a spatial dimension, the nature of their decision making is revealed by the patterns of land use and settlement recorded on plantation maps. The layout of plantation units tells much about the priorities of planters in terms of both economic and social control, so maps provide data on the character of everyday life in a form not duplicated by any other historical source. The map is not simply interchangeable with other types of communication, but provides unique information about perceptions and behavior.[2] Where a series of maps survives for a single plantation, even more can be learned about changing ideas and attitudes. This is not to say that maps by themselves are immediately explicable. As with any historical source material, the map reveals most when it is studied in conjunction with other types of evidence, used both as a check and as a means of understanding the significance of map data.

Purpose and Accuracy

Any assessment of the accuracy and quality of the cartographic work displayed by Jamaican estate plans must be framed in terms of the aims and limitations of the surveyors and planmakers. Almost all of the plans were commissioned for utilitarian purposes. They were expected to remain in private hands rather than being designed for publication. Certainly some plans were regarded as decorative objects, to be hung on the walls of absentees' English halls or in Jamaican great houses, and many exhibit sufficient craftsmanship and taste to take their place beside the portraiture, landscape art, and furniture of Georgian and early Victorian Jamaica.[3] Often the plans were simply nailed up in overseers' offices, however, regarded as heuristic devices rather than as objects of graphic art.

The utilitarian origins of the plans mean that a premium was placed on accuracy of representation and measurement. Planmakers and surveyors could expect continued employment only if they provided reliable data useful to their planter patrons. Although no systematic cartometric analysis has been attempted for the whole collection at the National Library of Jamaica, tests of point-to-point distance measures for a small number of examples suggest a range of error less than 2 percent for the gross dimensions of plantations.[4] This is a respectable level of overall reliability, but it does not mean that all elements within the plantation were measured with an equal degree of accuracy and care.[5] Above all, planters were concerned about boundary lines. These were confused considerably by failure to observe the true azimuth. The concern of this study, however, is strictly with those plans that show the internal layout of plantations, and the precise orientation of the plan matters relatively little for an analysis of internal land-use patterns so long as survey and protraction were consistent. Planters paid the additional cost of surveying each parcel of land within their plantations only because they wanted precise information for land-use management, and they expected a degree of accuracy similar to that obtained for boundary lines.

Certain features of plantation layout were more important than others in the planters' land-use planning and hence were treated differently by surveyors and planmakers. Fields planted in major export crops needed to be measured and plotted more carefully than areas of woodland or ruinate, for example, since the areas calculated for the former affected estimates of yield, productivity, and labor cost. Land planted in provision crops by slaves in "Negro grounds," on the other hand, was of less concern to the planter and so required less accurate plotting. The same principle applied to plantation buildings; works were the first buildings to be measured carefully and plotted in plan, while precise measurement of the location and dimensions of slaves' houses and even great houses was of less concern, and these buildings were portrayed in perspective for a longer period. These examples are sufficient to show that a blanket assessment of the accuracy of plantation maps is not very useful. Rather, it is necessary to consider the purpose of the plan in regard to each of its elements and to recognize that precision in one set of features was not necessarily matched by precision in all others.

Four principal elements of Jamaican plantation life in the eighteenth and nineteenth centuries will be discussed; plantation layout, works, houses, and gardens and grounds.[6] The emphasis is on an assessment of the value of the cartographic evidence for answering questions about the spatial structure of these elements, rather than geographical variation or change over time. But it will be useful first to offer a little more detail on the history of Jamaican land-use patterns during the period, in order to make the landscape more familiar.

Jamaica's Agricultural Landscape

By 1700 sugar had emerged as Jamaica's most popular crop, and by 1730 the island was firmly established as the British Empire's major producer.[7] In 1805 Jamaica produced almost 100,000 tons of sugar, the maximum for the entire period 1700–1900. In that year Jamaica was the leading individual sugar exporter in the world. The last decades of slavery were marked by gradual decline, while emancipation in 1838 was followed by a much more rapid contraction. Equalization of the British sugar duties between 1847 and 1854, removing the protection previously given British colonial production, followed by the removal of all duties on sugar imports in 1874, created further crises and decline.[8] This fall in sugar output during the late nineteenth century was cushioned to some extent by a shift to rum production, particularly on the north coast of the island. In summary, the eighteenth century was a period of substantial expansion for the Jamaican sugar industry, while the nineteenth was characterized by stagnation and decline.

No crop rivalled sugar as the leading export crop of Jamaica until the end of the nineteenth century. Coffee emerged as an important crop only in the 1790s, when it was granted British tariff protection and French planters fled to Jamaica from St. Domingue (including in their number the surveyor Jean Pechon). Production expanded rapidly, to reach an early peak in 1814 but, like sugar, stagnated until emancipation and then fell away rapidly until 1850. Unlike sugar, however, coffee experienced a revival in the second half of the nineteenth century. By 1890 coffee accounted for 18 percent of the value of Jamaica's exports, compared to only 15 percent for sugar.[9] The difference between the two crops was that sugar was not produced by smallholders on any scale, while coffee more readily made the transition from plantation to peasant production.

Pimento, ginger, cotton, indigo, and logwood were of no more than minor significance during the period of slavery. At the time of emancipation in 1838, pimento accounted for only 4 percent of Jamaica's exports and logwood for 2 percent, while sugar and rum made up 75 percent of the total. By 1890, however, logwood accounted for 21 percent, bananas 19 percent, pimento 4 percent, oranges 3 percent, ginger 2 percent, and coconuts 2 percent.[10] All of these were essentially peasant crops, though bananas were already being adopted by the plantation system and much of the logwood was cut from old plantation lands.

Alongside the dominance of sugar and other export crops, Jamaica always maintained a relatively significant internal trade and, in consequence, an economy much more strongly diversified than that of the sugar monoculture of the eastern Caribbean. The major reasons for this contrast may be found in Jamaica's size and the variety of its natural environment. The most important aspects of the internal market were the demand for livestock on sugar estates and the general urban and rural demand for foodstuffs. Livestock for motive power and meat were produced on a large scale on lands unsuited to sugar and coffee, and cattle competed with cane in some lowland regions. Food crops were produced by the slaves to supply both themselves and the free population, and later by peasants and wage laborers, using lands too rugged for sugar cultivation. The production of food crops took place either within the plantation system or on independent smallholdings, but livestock production took place on "pens" that often rivalled the plantations in area and scale of operations. At the time of emancipation these pens accounted for roughly 10 percent of the total value of Jamaican output.[11] After emancipation many sugar estates were converted to pens, so that their importance increased rather than waned.

Changes in the structure of Jamaica's economy between 1700 and 1900 were matched by changes in the spatial distribution of plantation production. English settlement was concentrated initially in the south and east of the island. Between 1740 and 1790 sugar plantations spread rapidly along the north coast and into the Westmoreland Plain, in response to the buoyancy of the sugar market and the availability of slave labor. James Robertson's map of Jamaica published in 1804 achieved a high level of accuracy and is particularly valuable since it portrays the distribution of sugar cultivation at the time of peak production (figure 6.1). By 1804 the density of "estates" (as all sugar plantations were known) was as great in the limited northern and western coastal plains as on the more extensive southern lowlands, except that the narrow, wet coastal fringe of Portland was sparsely settled and already in decline. Robertson showed a total of 830 sugar estates, but the number dropped to about 330 in 1854, and 125 in 1900.[12] By 1900 only a handful of mills continued to operate in the eastern half of the island, though the rum-producing estates of the north coast remained fairly dense and the Westmoreland Plain had just begun the process of centralization and amalgamation.

Coffee, grown on "plantations," rarely competed with cane but occupied distinct upland regions of Jamaica. The Blue Mountains produced coffee of superior quality, while the interior ridges

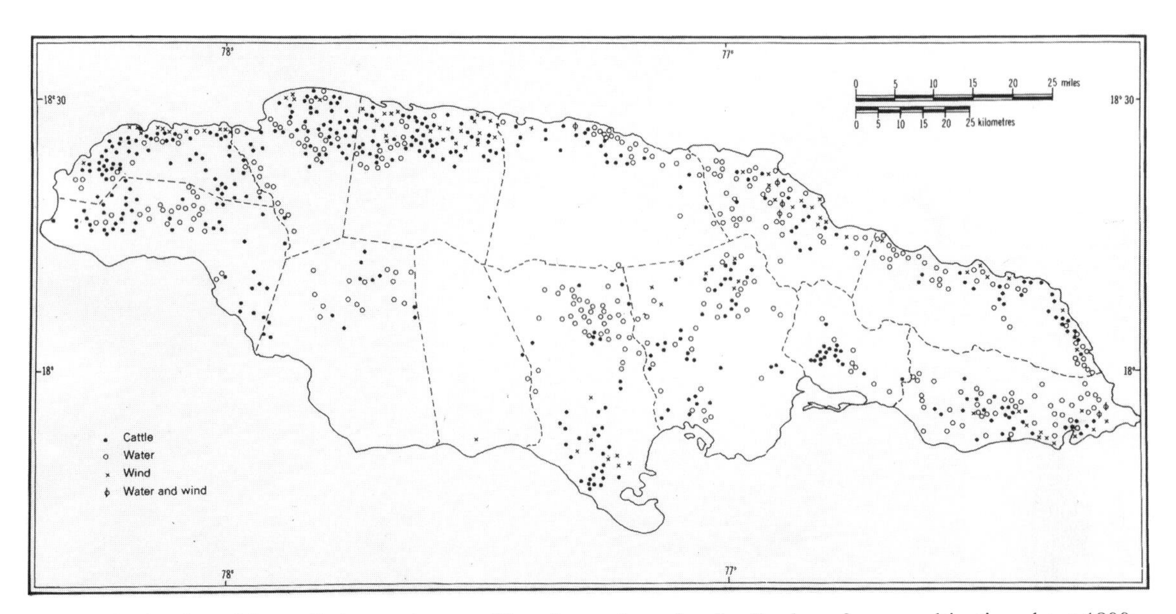

Fig. 6.1 Redrawing of James Robertson's map of Jamaica to show the distribution of sugar cultivation about 1800

of St. Catherine and St. Mary were also important after 1790. By 1799 there were 515 coffee plantations, with the focus in St. Mary.[13] The parish of Manchester, created in 1814, was for a time almost exclusively a coffee producer. Cultivation of steep slopes in the eastern parishes very quickly resulted in extensive soil erosion, and the rate of abandonment after emancipation was even greater than in the case of sugar. By 1900 there were only 33 plantations cultivating more than 50 acres of coffee, most of them in the mountains of St. Andrew and St. Thomas, whereas scattered small settlers cultivated almost 80 percent of the island's acreage. Only in the Blue Mountains did large plantations producing coffee of exceptional quality continue to dominate.[14]

Whereas the increasing dominance of sugar stifled the production of competing staple crops other than coffee, the growing demand for livestock on sugar estates was met by the increasing number of pens. In lowland regions, pens and estates competed for land space, and here pens were confined largely to marginal backlands and relatively infertile hilly areas. Where land of this type was limited, as in Trelawny and Portland, pens were few, and livestock was purchased from other parishes. Elsewhere, some specialized livestock-producing regions emerged, particularly in St. Elizabeth and St. Ann. Much of the land taken out of sugar after 1790 was put under grass, and penkeeping approached its maximum extent around 1900.[15]

Plantation Layout

The primary objective of any estate plan—beyond the mere glorification of the proprietor—was to provide information regarding the spatial layout of the plantation. In general, the information offered by Jamaican plantation maps toward this end was of high quality (see plate 8). But patterns of plantation layout varied in response to a variety of factors, some of them affecting modes of cartographic representation in a direct fashion. Most important were the principal crop and physical setting of the plantation, and these two features were interdependent. In order to interpret the land-use patterns shown on Jamaican plantation maps, then, it is necessary to relate them to the physical environment. As noted in chapter 5, only a small proportion of the plans showing internal layout provide data on variations in relief, and none of them contain information on climate or soils. In this chapter a smaller sample of relatively complete plans is drawn from the 856 discussed earlier, and these sample plans are related to a range of physical and locational data. This sample is used to derive an understanding of the general principles underlying plantation layout, while the range of deviation from these tendencies is illustrated by discussion of particular examples.

All of the 856 plans from the National Library collection show some feature of the internal layout of Jamaican plantations, but many are incomplete in one respect or another. The sample studied here consists only of those plans that show the internal layout of plantations in their entirety, possess full reference tables, and can be dated precisely. Plans have been excluded from the sample for a number of reasons. Some are damaged or illegible. Others show only one element of plantation layout, such as works buildings or cane pieces (fields). Some plans detail field patterns throughout but provide only partial reference tables, so that land use cannot be established certainly. A few plans cannot be located precisely. Plans lacking scales have also been excluded, though comparison with modern topographic maps might permit the calculation of scale at an acceptable level of accuracy for a good proportion of these and so increase the number of usable plans. Plans lacking dates have been included only if they can be placed within a range of five years on the basis of internal evidence. Copies have been excluded from the sample even when they pass all other tests, since they contain elements of later cartographic style and potentially distorted land-use data.

Only 282 (33 percent) of the 856 plantation maps satisfy these rigorous conditions, but this is a substantial sample and provides an adequate basis for analysis. In certain respects, the sample is more representative of the plantation system than is the total collection. Some 47 percent of the sample plans relate to sugar estates compared to 52 percent of the total, while 31 percent of the sample relate to pens (22 percent of the total); but only 22 percent of the sample plans are for plantations, compared to 26 percent in the total collection. This does something to redress the weighting toward sugar. There is no significant difference between the sample and the total collection in terms of the distribution of plans over time, except that the peak around emancipation is even more exaggerated for the sample. Some 20 percent of the sample plans relate to the decade 1830–39, compared to 13 percent of the total, but the sample shows relatively less surveying activity in the 1820s and 1840s. In terms of distribution in space, the southern parishes stretching from Manchester to St. Andrew are underrepresented. These tendencies help to even out some of the inequalities in the total collection, though the importance of St. Andrew, with 11 percent of the plans in the sample, is inflated along with the number of pens. In general, however, the sample reflects closely the structure of the total collection and provides a fairly representative cross-section of Jamaican plantation history between 1700 and 1900.

Information relating to the number of field units devoted to each type of land use and their acreage can be obtained directly from the reference tables of the plantation maps. In addition, a range of measurements can be taken from the plans, relating to the location of fields and other land-use elements. These measurements are shown in figure 6.2. Most of them have their origin in the works of the plantation, including under this designation sugar mills, coffee pulping houses, and cattle pens. Sugar estates rarely had more than one mill location, but coffee plantations sometimes had several sets of barbecues and pulping houses, and pens frequently had more than one cattle pen (meaning a small walled or fenced enclosure rather than the whole property). Where multiple works locations occurred, the largest was used, and where any uncertainty remains, the location nearest the great house was selected. The measurements are as follows; midpoint of works to most distant property boundary (*B*), to nearest boundary (*C*), to midpoint of farthest field under the main crop (*D*), to midpoint of farthest field under pasture (*E*), to midpoint of farthest provision ground (*F*), to midpoint of workers' village (*G*), and to great house (*H*). The distances from the workers' village to the great house (*I*) and to the provision grounds (*J*) were also measured. Indexes of the shape of the plantation as a whole were calculated from measurement of the perimeter (*A*) and longest axis (*K*), together with area.[16] Field shape was not measured precisely, but a subjective classification was made, distinguishing square, rectangular, elongated, straight-sided but irregular, and sinuous irregular types.

In addition to these data derived directly from the plantation maps, a number of physical environmental indexes were calculated. Distance to the coast provides an index of isolation. Rainfall, maximum altitude, relative relief (the difference be-

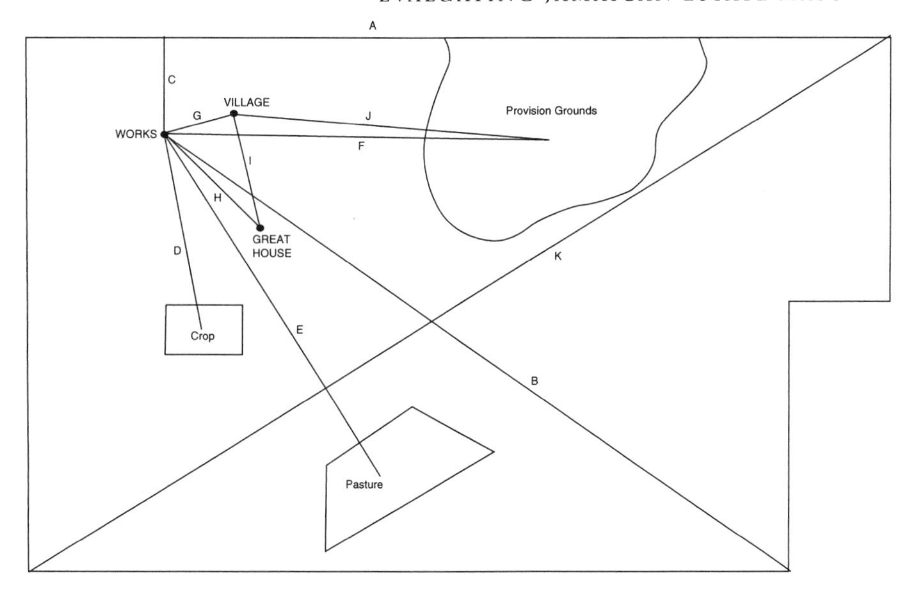

Fig. 6.2 Measurements relating to the fields and other land-use elements on the plans

tween the highest and lowest point), and average slope were established for each of the 3-kilometer-square quadrants of the 1:50,000 topographic map, and these are taken to be typical of the plantations located within them.[17]

Analysis of these variables permits a full study of the changing size and shape of Jamaican estates, plantations, and pens, and of the central location of their mills. Sugar planters placed great weight on the economic advantages of locating mills centrally, to minimize the costs of carting cane and other goods, but in practice a large proportion broke this general rule. The reasons for the failure to find central locations were chiefly topographic and related to the availability of power sources, particularly water. Sugar estates settled on the southern plains of Jamaica in the early phase of expansion were most likely to be compact in shape and homogeneous in terms of soil and slope. Such estates usually had a large proportion of their area under cane, so that a works site central to the cane fields was also central to the whole estate. For example, James Cradock's 1758 plan of Parnassus Estate on the Clarendon Plains shows a high index of centrality for the works (plate 5). The works on this estate were animal-powered, but another of the Dawkins plantations, Friendship, located in the more rugged land of upper Clarendon, achieved a lower index by carrying water through a gutter to drive the mill. The water was taken from a river and a spring, crossing streams by means of viaducts, and stored in a dam to ensure constant supply. In some cases water was even carried from points outside estates to centrally located mills. Local variations in topography sometimes prevented central location of the works, however, even when water was taken off a stream and carried in a gutter for a considerable distance. Such a situation occurred at Fort Stewart in St. George, surveyed by Murdoch and Neilson in 1795 (figure 6.3). There the water-powered mill was sited outside the cane fields, on the far side of the Back River. These anomalous examples all suggest that a water-powered mill was more to be desired than a central location for the works,

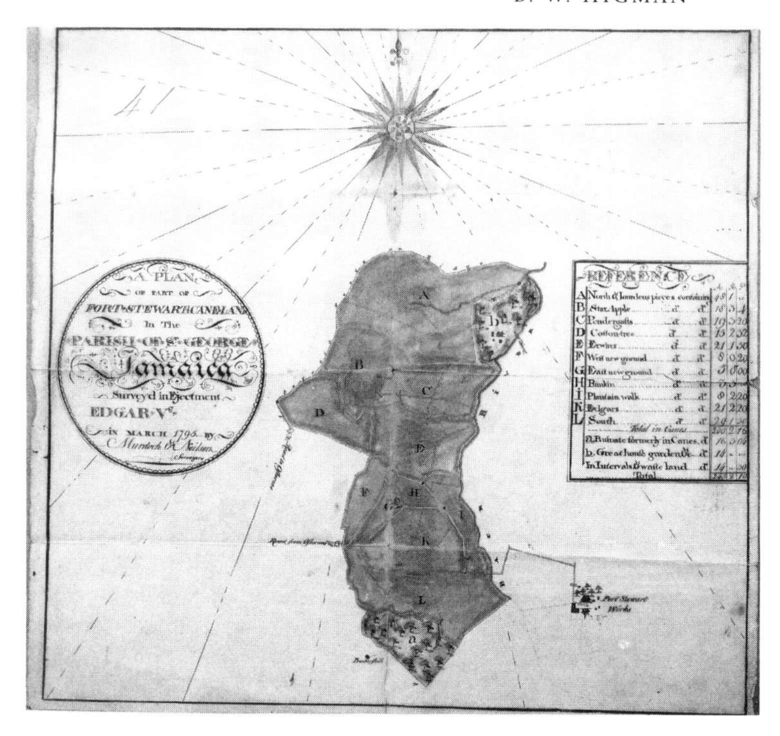

Fig. 6.3 Murdoch and Neilson, plan of part of Fort Stewart Cane Land, 1795 (Courtesy of the National Library of Jamaica)

though the planter sought to gain both ends wherever possible. Failure to site sugar works at the very center of estates reflected a series of compromises between the desire to follow ideal models and the constraints imposed by the shape, topography, soil, and power resources of plantations.

The relationship between topography and estate layout can be understood more fully by superimposing contours on plantation maps. Two examples follow.

Orange Cove Estate in Hanover was surveyed by Philip A. Morris in 1816, but his plan survives only in the form of a later tracing. The outlines of this plan are matched to the coastline and contours in figure 6.4. Although some small variations occur in the coastline according to Morris and the modern topographic sheet, these may largely be attributed to actual coastline change, and the match between the 1816 and modern road systems is very close indeed. In 1816 Orange Cove was owned by William Allen, an absentee, who died about 1830. The 260 slaves laboring on the estate produced some 250 hogsheads of sugar and 100 puncheons of rum. In 1804 the estate had a windmill and a cattle mill, and by 1880 it had a steam mill and centrifugal. In 1816 some 261 of the estate's 430 acres were in cane, but in 1880, when Orange Cove encompassed the neighboring estate Esher, only 185 acres were in cane and output had dropped to 190 hogsheads of sugar.[18] The works were located roughly midway between the eastern and western boundaries of the estate, but pulled north of center by the need to catch the wind and by the attraction of the wharf on the western side of the mouth of the bay whence the produce was shipped to England. The great house, occupied only by an overseer in 1816, was

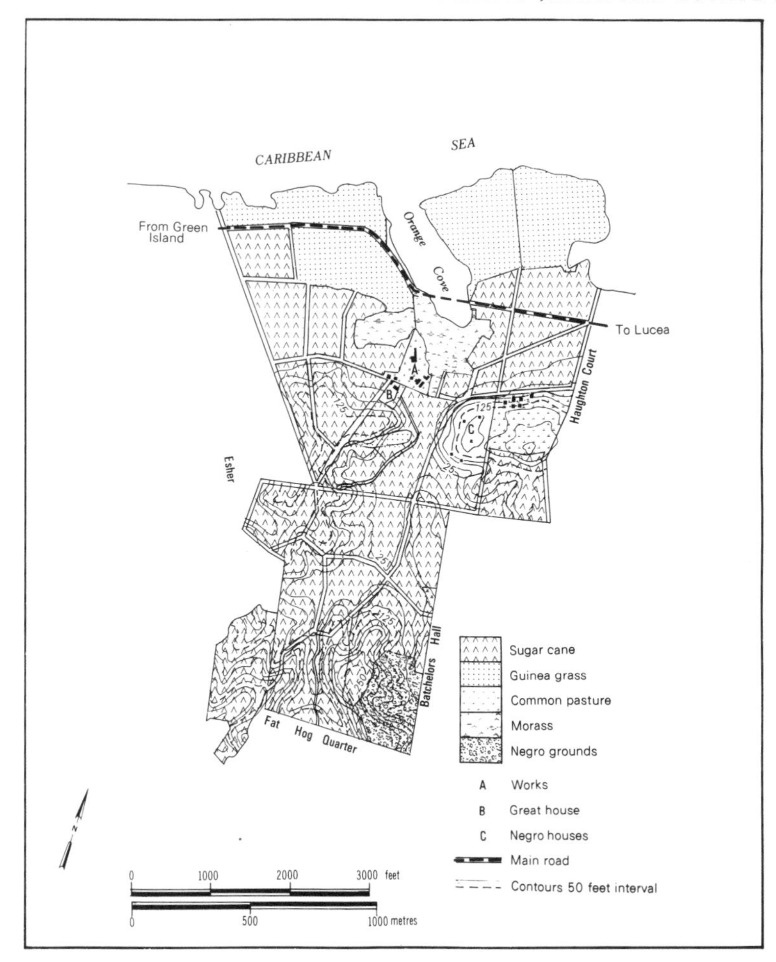

Fig. 6.4 Redrawing of Philip A. Morris's plan of Orange Cove Estate, 1816

immediately opposite the works. The "Negroe houses" occupied 11 acres of relatively hilly land to the east of the works, and some houses were also strung along the road beside the common pasture field called "New Negroe house." The estate was divided among three ecological zones: a level coastal fringe; low hills surrounding flatlands, south of the works; and steep hills at the southern margin. Cane was grown in each of these zones, except that guinea grass occupied the brackish lands nearest the sea, taking up 87 acres, and the steepest hills contained the meager 14 acres of provision grounds for the slaves. The 19 acres of common pasture was divided between the 4 acres surrounding the works and the steeper hill to the east. Field shape at Orange Cove varied in response to differences in the environment, taking straight lines in the level lands and twisting around the bottoms of hills and edges of the morass.

Wheelerfield Estate was one of the series of estates fronting the Plantain Garden River in St. Thomas. The outlines and land use of the estate are known from a plan of 1858 by an unnamed surveyor, and these have been matched with the contours in figure 6.5. With a total area of 910 acres, Wheelerfield was more than twice the size of Orange Cove but stretched through a similar range of ecological zones. Wheelerfield possessed much more steep land, with a relative relief of 1,000 feet compared to the 250 feet at Orange Cove. Thus the pattern of land use was more clearly demarcated at Wheelerfield. Cane extended from the river to the 75-foot contour line, with the margin of cultivation following that line almost exactly, and cane was the only land use within that zone except for pasture close to the works and on the poorer, flooded soils of the river's slip-off slope. An appreciation of the importance of this line can be seen in the plan of 1858, which notes the "foot of the hill" at this point. Pasture and ruinate occupied the lower slopes in the northeastern section of the estate, while woodland covered the steepest hills to the

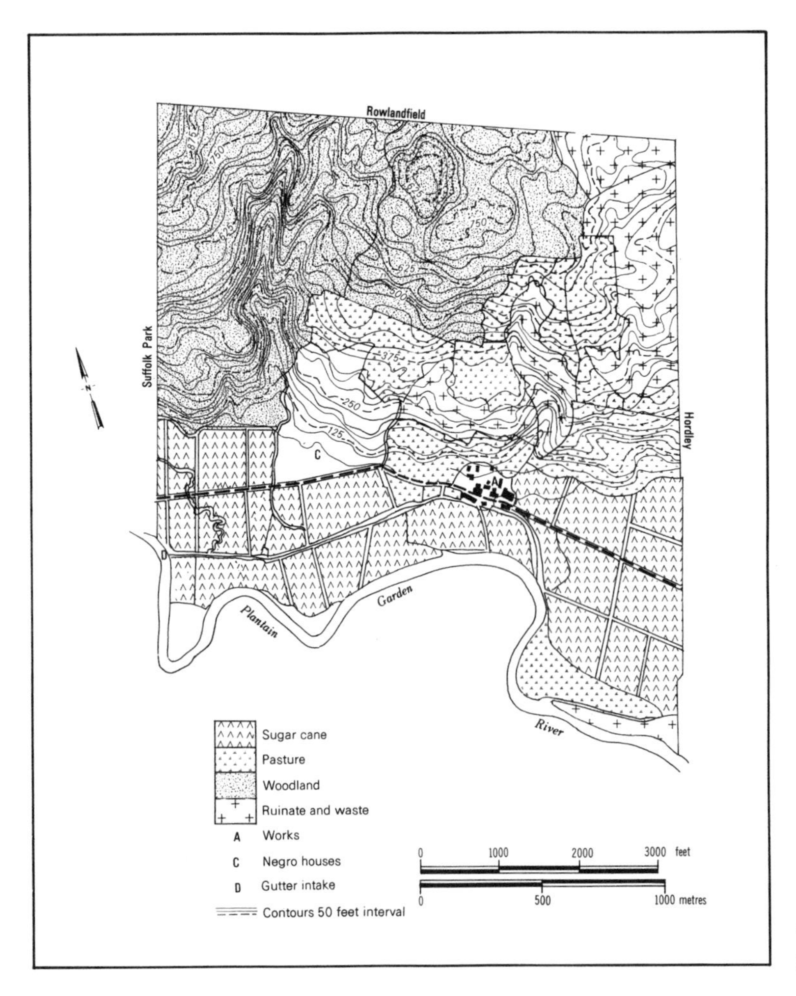

Fig. 6.5 Redrawing of an anonymous plan of Wheelerfield Estate, 1858

northwest. Robertson's map of 1804 shows that Wheelerfield depended on a water mill, and this remained the state of technology until 1880 at least. In 1832 the estate was worked by 322 slaves, who produced 233 hogsheads of sugar and 177 puncheons of rum for the absentee owner Thomas Milner. In 1858 some 210 acres were in cane, but by 1880, when the property belonged to the Colonial Company of England, only 120 acres were in cane, and the estate produced only 112 hogsheads of sugar and 104 puncheons of rum.[19] The works were located near the center of the cane pieces, while the "Negro houses" occupied 42 acres on the lower slopes of the woodland zone. The cane fields took regular geometric shapes except where they bordered the river or the hills, while the pastures generally followed the natural contours.

The deviations from the ideal model apparent in the layout of Orange Cove and Wheelerfield Estates are explicable principally in topographic terms, but they resulted also from economies of scale, the need to provide power inputs with differing technological requirements, and access to major transportation routes. Thus cane was pushed into more distant, steeper hills at Orange Cove as a result of the relatively small size of the estate. At Wheelerfield the layout approached the ideal model more closely, except that the river cut a segment from the concentric circles constrained by the estate's rectangular shape. In general, then, differences in the layout of sugar estates can be explained by factors of ecology, shape, and scale, all of which operated to prevent planters from imposing any model on the landscape regardless and created varied spatial frameworks for the development of plantation society.

A central location for the complex of works, village, and great house was seen as being just as important for a coffee plantation as for a sugar estate. In the 1790s P. J. Laborie, another St. Domingue émigré, advised that "[t]o fix upon the centre of the tenement is a very material precaution, chiefly if the estate is extensive and the lands are not of durable quality."[20] If a public road passed through the plantation, however, it was wise to site the "settlement" at a distance so that the planter would not be disturbed by travellers or "the interior order and discipline of his negroes" upset. Again, the center of the plantation might be excessively steep, or lack water, timber, and stone. But Laborie concluded that, "where every necessary thing does not lie contiguous, I would prefer to give up those accessory conveniences, rather than abandon the centre, especially if water can be conducted thereto through a pipe. The establishment is permanent, and its situation determines for life the convenience and easiness of every future service; and the fatigue of ordinary labour is much increased, when daily performed at a great distance" (13). This advice was very similar to that given to sugar planters. Although the works required for coffee were less complex and industrialized, a mill was necessary to pulp the berries and peel the beans, and it was more efficient if water-powered than if driven by animals.

The works on coffee plantations occupied smaller areas than the factories of sugar estates. Whereas the works of a sugar estate were proportioned to the total size of the property and the area planted in cane, the area of coffee works varied independent of either of these factors and indeed of all elements of plantation land use. The centrality of the works locations on coffee plantations was not significantly correlated with plantation size or shape, or with any of the environmental characteristics included in the analysis. Thus variations in the location of coffee works must be traced to local circumstances affecting particular plantations.

The shifting character of coffee cultivation meant land use and field patterns were constantly changing, so that any plantation map provides only a single picture, cutting through the process in an arbitrary fashion. Compared to the pattern on sugar estates, a relatively small proportion of coffee plantation land was in cultivation at any one time. Only 14 percent of the land was in coffee on the sample of plantations for which plans are available over the period 1780–1859, compared to the 24 percent of sugar estate land in cane. A similar contrast between coffee and sugar occurred in the proportion of land devoted to guinea grass and common pasture. Coffee plantations, especially those in the eastern region, always had large proportions of their area in woodland, provision grounds, and, especially after 1820, ruinate. These uses took up 58 percent of the land on average, compared to only 43 percent on sugar estates. The area allocated to provision grounds was more generous than on sugar estates, but plantain walks (groves) disappeared around 1800 as slaves were increasingly forced to fend for themselves. Variations in land-use patterns on coffee plantations were related chiefly to differences in the physical environment. Thus the area planted in coffee was not proportioned to the total area of plantations but increased with altitude.

Coffee plantations were divided into fewer and larger field units than the typical sugar estate. In the sample of plans studied here, the mean number of total field units on coffee plantations was 19.2 (standard deviation = 10.4) for the whole period 1790–1859, compared to 44.8 on sugar estates. Mean field size was 46.6 acres (SD = 113.2) on coffee plantations, but only 25.7 acres on estates. This contrast is explained by the relatively large proportions of wood, ruinate, and provision grounds found on coffee plantations, which were commonly divided only into gross areas. The size of cane and coffee pieces differed much less dramatically than the overall comparison might suggest. Thus the mean area of coffee pieces was 10.9 acres (SD = 6.4), little more than the 9.7 acres typical of cane pieces. Coffee plantations simply had fewer pieces. Whereas the typical estate had 26.8 cane pieces, the mean number of coffee pieces on a plantation was only 10.9 (SD = 8.3).

It is necessary to enter a caution at this point. The coffee pieces discussed above consist of the named units called pieces on plantation maps, but these were sometimes divided into smaller numbered units. For example, Thomas Harrison's 1863 plan of Tweedside Plantation, high in the mountains of St. Andrew, showed 24 coffee pieces separately named and lettered *A* to *X* (figure 6.6). Each of these pieces was given a distinct color on the plan. Some of the pieces were divided into as many as six numbered units, but the numbers appeared only in the reference table and were not shown on the plan itself. In the case of Rampie piece (*X*) the numbered units were described as containing "young coffee," or being "partly planted" or "not yet planted," but elsewhere the logic of the numbering is not clear. In some cases the numbered units were fragmented, but the most fragmented pieces were not divided into numbered units. The importance of this point is that, while Tweedside contained 24 coffee pieces with a mean size of 4.3 acres, there were 52 numbered subdivisions with a mean area of only 2.0 acres. It appears, then, that coffee planters used named pieces to refer to the major units of cultivation but broke these down further for accounting purposes on the basis of stages of growth. Sugar planters rarely divided cane areas below the piece and never planted canes in the tiny, fragmented plots common on coffee plantations. In large measure this contrast reflected the different topographic conditions of sugar and coffee

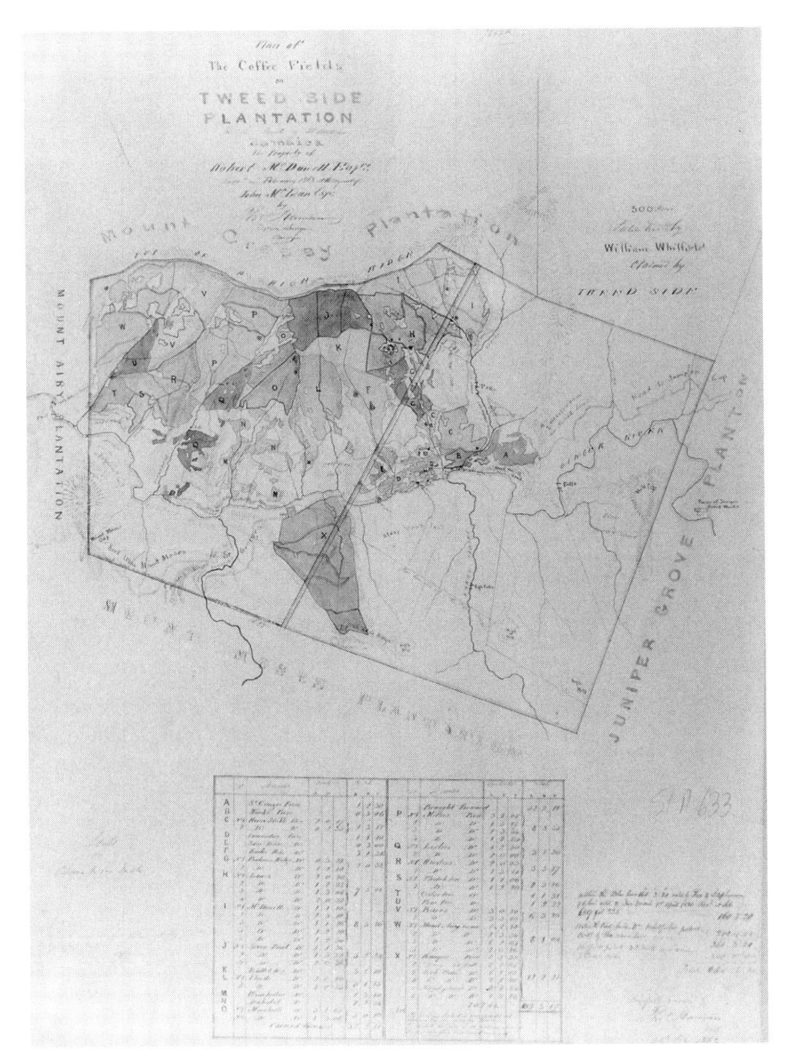

Fig. 6.6 Thomas Harrison, *Plan of the Coffee Fields on Tweedside Plantation*, 1863 (Courtesy of the National Library of Jamaica)

culture. Since most plantation maps did not show these small, numbered units, the analysis here is based strictly on pieces, but the alternative results that might emerge must be borne in mind.

Differences between the field patterns of sugar estates and coffee plantations were most pronounced in terms of shape. Only on 5 percent of plantations were the coffee pieces made up of regular geometric figures, compared to 30 percent of cane pieces on estates. Some 82 percent of coffee pieces were irregular units with curved, natural boundaries, compared to only 36 percent of cane pieces. Regular figures occurred only in areas with slopes of less than 15 degrees, and below 3,000 feet. This close relationship between field shape and topography, for coffee and pasture, created a strong regional contrast between plantation layout in Manchester and the Blue Mountains.

Waltham Plantation was located in Manchester on the southern fringe of the growing town of Mandeville. The land within this area was a plateau of gently rolling hills, ranging between 2,000 and 2,250 feet above sea level and merging into steeper slopes at the margins. The field pattern at Waltham followed the contour lines to some extent, but contained a good proportion of straight-sided figures. James Windett's plan based on a survey of 1830 showed that 207 of the plantation's 650 acres were in coffee, but by 1847 coffee had been abandoned (figure 6.7). The property was worked by 155 slaves in 1832 for an absentee owner. The coffee pieces were sprinkled with guinea grass, common pasture, and even woodland forming an intermediate zone around the works, and the provision grounds and coffee cultivation were juxtaposed.[21]

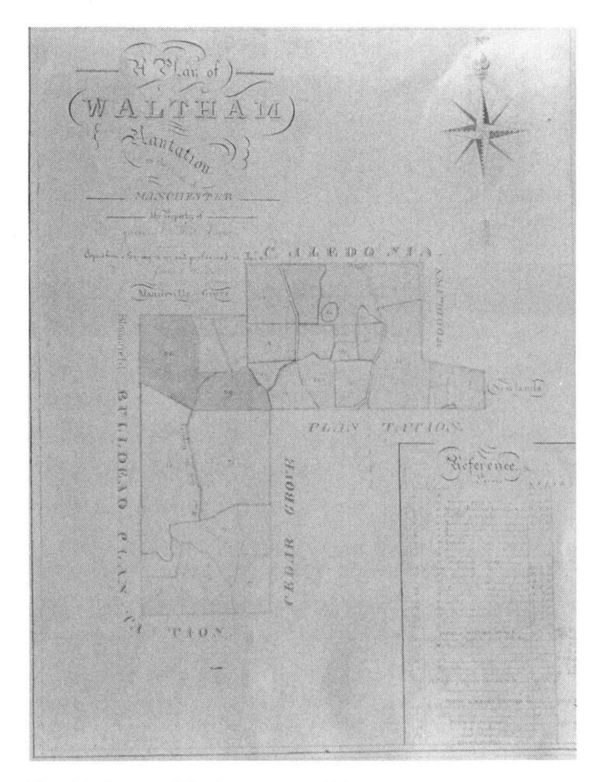

Fig. 6.7 James Windett, plan of Waltham Plantation, [1830] (Courtesy of the National Library of Jamaica)

Tweedside Plantation provides a graphic illustration of the much closer matching of field patterns and topography that occurred in the steep slopes of the Blue Mountains (figure 6.6). Thomas Harrison's 1863 plan of Tweedside was highly successful in demonstrating the relationship between ridges, spurs, drainage networks, and fragmented coffee fields. Other plans show that the pattern of dispersed coffee pieces, beyond a zone of pasture surrounding the works, occurred in the less rugged areas of the eastern coffee region as well as in Manchester. Thus the field pattern found high in the Blue Mountains may be seen as a unique response to the high costs of movement in such rugged country, balanced against the high value of the product.

The pattern of land use on Jamaican pens was relatively simple. The pens lacked the elaborate works of the sugar estates and coffee plantations, and cultivated crops only as a subsidiary to grass and livestock production. These supplementary crops such as pimento, coffee, and cotton were generally identified on the plans in the normal way. It is instructive, however, to notice an important cartographic exception. Logwood, used in the production of dye, rarely appeared on plantation maps, though other historical evidence shows it to have been an important source of income in some periods. The reason for this omission was chiefly that logwood was not planted systematically and was often used only to form hedges or fences. Introduced from Honduras in 1715, logwood was self-sown and came to overrun extensive areas by the 1780s, particularly in St. Elizabeth and Westmoreland. It was exploited most intensively in times of hardship, notably the later nineteenth century, when other products were relatively unprofitable. The only plans in the sample to show significant areas in logwood relate to the 1860s. John Manderson's 1864 plan of Fullerswood Pen, near Black River in St. Elizabeth, listed nearly 1,200 acres of "ruinate and logwood" in the pen's total area of 2,146 acres. Earlier, in 1832, Fullerswood Pen ran 370 head of livestock and sold working cattle to estates and cattle and sheep to the Black River butchery. In the same year it shipped 100 tons of logwood.[22] It seems probable that many other pens and estates had productive logwood among their ruinate and forming their fences, and this is one of the few respects in which the plans fail to provide a "true" picture of land use.

The field units of pens were larger than those typical of sugar estates but smaller than those of coffee plantations. The shape of grass fields was more regular on pens than on estates and plantations. They were also more regular than coffee pieces but less geometric than cane pieces. This pattern reflected differences in field size together with the intermediate topographic conditions of pens. Pens with rectangular fields were all located in areas with mean slopes of less than 10 degrees, while the proportion with irregular, natural field shapes increased steadily from 21 percent below 5 degrees to 67 percent above 20 degrees. A similar relationship between field shape and altitude was evident, with a critical point reached at about 2,000 feet above sea level.

The relationship between field shape and topography may be illustrated by the example of Union Pen, located on the rolling uplands of St. Ann, close to the border with St. Mary. The layout of Union Pen is known from a plan of about 1820, by an unnamed surveyor, and has been reproduced in outline with the contour lines superimposed in figure 6.8. At the date of the plan, Union covered 1,130 acres, with 803 acres in guinea grass, 51 acres in common pasture, 31 acres in pimento, 35 acres

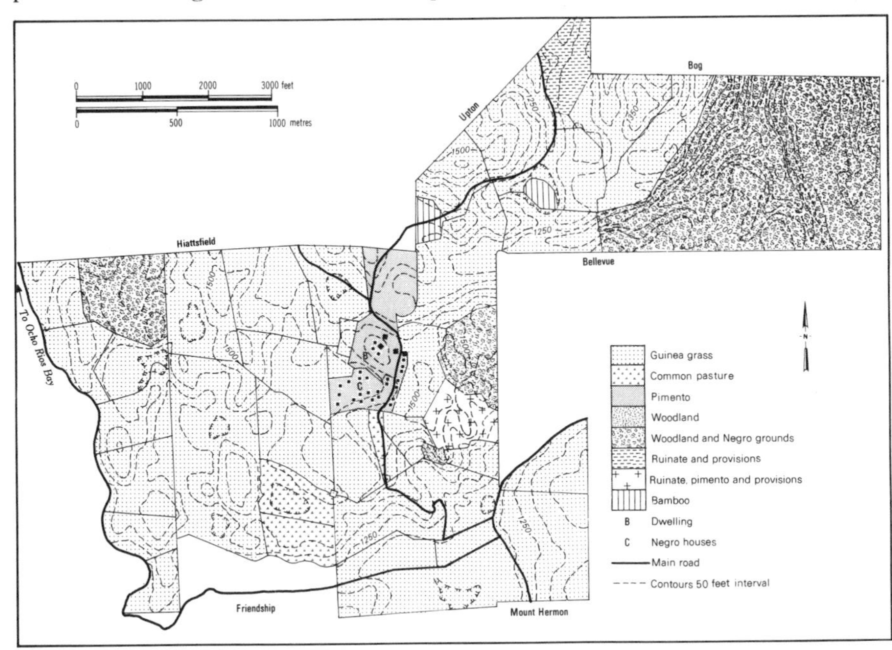

Fig. 6.8 Redrawing of an anonymous plan of Union Pen, c. 1820

in ruinate and provisions, and 204 acres in wood and Negro grounds. In 1832 there were 513 head of livestock on Union. It was worked by 122 slaves for the resident owner, a free colored man, Benjamin Scott Moncrieffe, who also acted as attorney for other properties. Moncrieffe appears to have established a butchery on Union and sold beef as well as working cattle to neighboring estates. He also owned race horses that won purses for him at meets in Montego Bay, Kingston, and Clarendon.[23] The pimento on Union was concentrated in a small zone close to the "dwelling house" and its barbecues, perched on a hilltop site and overlooking the slaves' houses. This central zone was surrounded by an expanse of guinea grass stretching to the pen's boundary fences except in the areas occupied by wood and provision grounds. Although the lands of the pen were by no means a uniform plain, the field boundaries generally followed straight lines, often proceeding right across the property and creating fields with irregular but straight-sided shapes. Only where the field boundaries coincided with public roads or met the steeper slopes of the hills in woodland did they hug the contours. The demands of grass in terms of soil and slope were much less confining than those of cane or coffee, of course, so it was easier for the penkeeper to impose a geometric pattern on the landscape.

This brief review of some aspects of plantation layout shows that map evidence has clear advantages and disadvantages. It provides unique information about the internal and external spatial organization of plantations, offering exact answers to questions discussed only loosely in impressionistic literary sources. Measurement permits precise generalization about the relative location of works, great house, and workers' village, the fundamental trinity of plantation layout, and enables comparison of the pattern on sugar estates, coffee plantations, and pens before and after emancipation (figure 6.9). The plans also provide data necessary to a study of land-use patterns, but it is important to match these data with information about production wherever possible. The nature of logwood and pimento production, for example, meant that they appeared as items in plantation account books but were not always indicated on plans.

Works

In contrast to the high level of accuracy and detail achieved by Jamaican surveyors and planmakers in their representation of plantation layout, the information provided with regard to works, houses, and gardens and grounds is meager. The plans are useful, certainly, in supplementing information from other sources, but systematic analysis would be pointless or impossible. This inadequacy is not a result of the limitations of contemporary surveying techniques. Rather, it is a product of the limited interests and demands of the planters who commissioned the work. Very often, the detailed data provided by unusually particular surveyors seem to have been supplied voluntarily rather than in response to any concern on the part of their planter patrons.

Plantation surveys frequently indicated the location of works buildings in perspective or plan but seldom identified the functions of particular buildings. In some cases the dimensions and locations of these buildings were measured and protracted precisely, as revealed by surveyors' fields books and the presence of pinpricks at the corners of the structures. Often, however, the buildings were merely sketched in by eye, particularly in surveys of the early eighteenth century. Planters required more detailed surveys only occasionally, and the number of surviving plans that show the internal

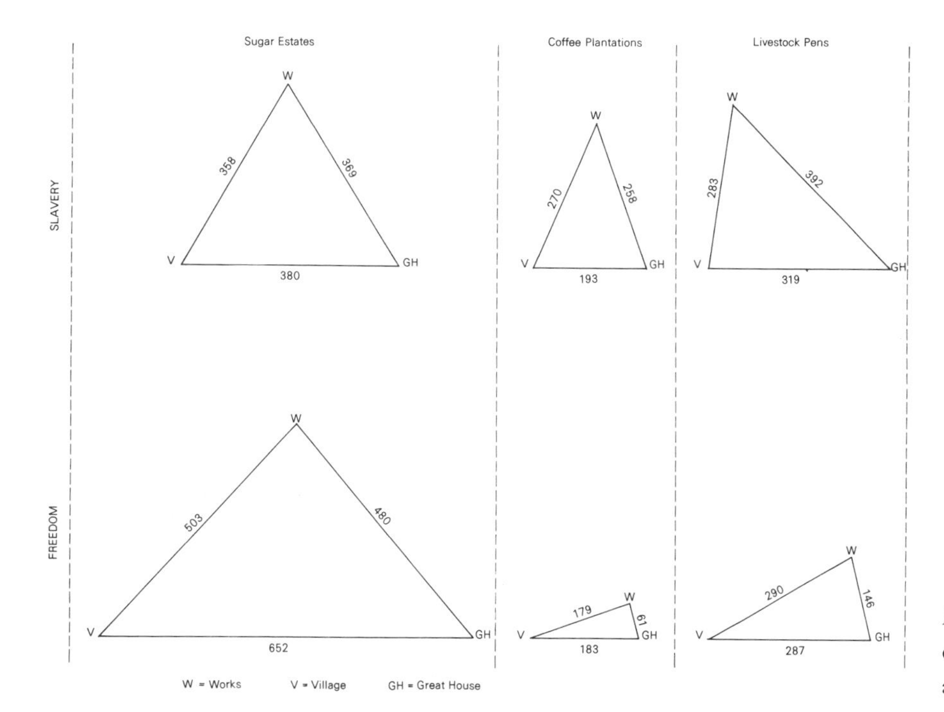

Fig. 6.9 Relative locations of chosen sites before and after emancipation

organization of works yards is small. All of these relate to works on sugar estates and coffee plantations, and the number of examples is too small to permit a systematic analysis.

Some general points can be made by looking at examples considered in chapter 5. Cradock's 1758 plan of Parnassus was precocious in showing the buildings of the estate's sugar mill in plan, and it could be that he indicated the shape and location of the buildings with precision (plate 5). But the key to the plan merely states that fields *a* and *b* contained "pasture about the works." The functions of the individual buildings can only be guessed at. The cane field labelled *K*, to the east (below the works), was called "Cureing house piece," but since field *K* bounded the entire length of the works yard, it is impossible to use this information to identify a particular building as the curing house.

Grant's 1782 plan of Mount Ephraim Estate (figure 5.6) employed a less sophisticated cartographic technique, showing the mill buildings in perspective and failing to identify the works area in the reference. (Interestingly, the "old works" in the common pasture to the west were partly drawn in plan.) But it is at least possible to establish from the perspective drawings that the mill used water power, probably an undershot wheel dependent on the flow of the river. The building immediately to the east of the mill must have been the boiling house, and the next building was probably the distillery, while that to the north of the mill was probably the trash house (used to dry crushed canes to fuel the coppers of the boiling house).

Infinitely superior to either of these two plans was Pechon's 1800 "geometrically performed" survey of Belvidere (figure 5.3). Not only is there good reason to believe the buildings were accurately measured and located, but Pechon also carefully identified the function of each in his numbered key. The buildings were drawn in plan with hipped roofs. The water mill (building 7) was fed by an aqueduct drawing water from a dam near the northern boundary of the estate. South of the mill were the T-shaped boiling and curing houses (5) and the still house (6) in the shape of a cross. Two trash houses (8) bordered the cattle pen (9) and mule stable (10). The other two buildings within the works yard, covering in all 8 acres, were the corn house (11) and the cooper's shop (12). The blacksmith's shop (13) was located a little beyond the works yard, to the east. Although Pechon provided relatively little by way of architectural detail, he offers almost everything else the historian might ask: information on the differentiation and specialization of processes, the flow of activity, the scale of operation, and the integration of support services. Unfortunately, few surveyors were as meticulous as Pechon, and cruder methods of representation remained common throughout the nineteenth century.

Houses

Jamaican surveyors and planmakers were, if anything, less careful in their representation of plantation houses than of works buildings. Whereas the rare detailed surveys of works were generally based on precise survey, plans of plantation workers' villages were often impressionistic. Great houses were large, individual units and so more easily plotted, but their precise dimensions were not always measured. Thus, while protractions of works often show the pricks of dividers at the corners of each building, the houses of slaves and wage laborers rarely show any pricks at all, and a prick in the center of a house represented a relatively great degree of care. Thus map evidence relating to the internal spatial structure of plantation villages must be treated with great care. Caution must also be exercised in looking at the evidence they offer on great houses and their outbuildings.

The great houses of Jamaican estates varied considerably in scale and architectural magnificence, but the plans do not provide systematic data. As for works, Pechon's 1800 plan of Belvidere offers the most useful kinds of information (figure 5.3). The great house (1) is shown perched on a hill overlooking the cane fields and works, the house itself being no larger than the boiling house, with a hipped roof. Behind the great house were the offices (3) and to the west the overseer's house, dry goods and provisions stores, stable, and coach house (2). Forming part of the same cluster were the hothouses (4) and the slave hospital.

The slaves' houses at Belvidere were located to the north, spread along the next ridge, close to the great house complex but clearly separated from it. Here Pechon's precision disappears, the slaves' houses drawn freehand and, apparently, without attention to their exact locations within the village area. Pechon indicated about seventy houses, of varying size, and this may have been near to correct. The slave population of Belvidere was then approximately 350, and five to a house was typical. Pechon also surrounded groups of houses with fences, suggesting compoundlike household organization, and here again the evidence indicates that this was a realistic picture of the Jamaican slave community on large sugar estates.

While Pechon's plan of the slaves' houses on Belvidere was less than precise, and lacks the care obviously taken in his survey of the works and great house complex, it is better than most plans of Jamaican plantation village sites. Only a handful provided genuinely accurate surveys, almost all of these coming after emancipation.[24] In spite of their general lack of accuracy, Jamaican estate plans constitute valuable sources of information about the layout and architecture of workers' houses both before and after emancipation. The value of the information they supply lies not in its precision or comprehensiveness but in its uniqueness. The reluctance of landscape artists to depict the houses of plantation workers was shared by later photographers, and literary accounts are often vague and impressionistic. In part this attitude stemmed from the perception of the plantation village as almost separate territory. Many whites never entered the houses of slaves because they "considered them as secluded," and slaves and apprentices, in turn, attempted to protect the privacy of the village community.[25]

Land surveyors, on the other hand, were required at least to traverse the boundaries of the workers' villages, and this put them in a position to make close observations of the layout of the houses. The accuracy of such observation depended on the size of the village area, its topography, and the extent to which the houses were enveloped by vegetation. If the plan was produced from field notes by a person not actually involved in the survey, all this observation might not be reflected in the final map, but this was rarely the case. Certainly two versions of a plan might show different patterns of layout, reflecting the preconceptions of the planmakers. For example, Edward McGeachy was commissioned to make a survey of Warwick Castle Estate in St. Mary but discontinued his work early in 1841 when the attorneyship of the estate was changed. His plan has a blank at the site of the workers' houses. In June 1841 Richard and Robert M. Wilson surveyed the entire estate. One version of their plan, which is unsigned, shows eighteen workers' houses, regularly distributed along horizontal lines and their doors facing west (figure 6.10). Trees are scattered among the houses, and a pond is shown. A second version of the plan, probably by a different hand but of the same scale and matching the first exactly in outline, shows only fifteen houses, scattered irregularly and concentrated toward the eastern side of the area, their doors facing east (figure 6.11). No pond is shown in the second version, though the path through the village follows the same route on each plan. The standardized architecture of the houses is the same on the two plans, but may indicate simply a symbolic house,

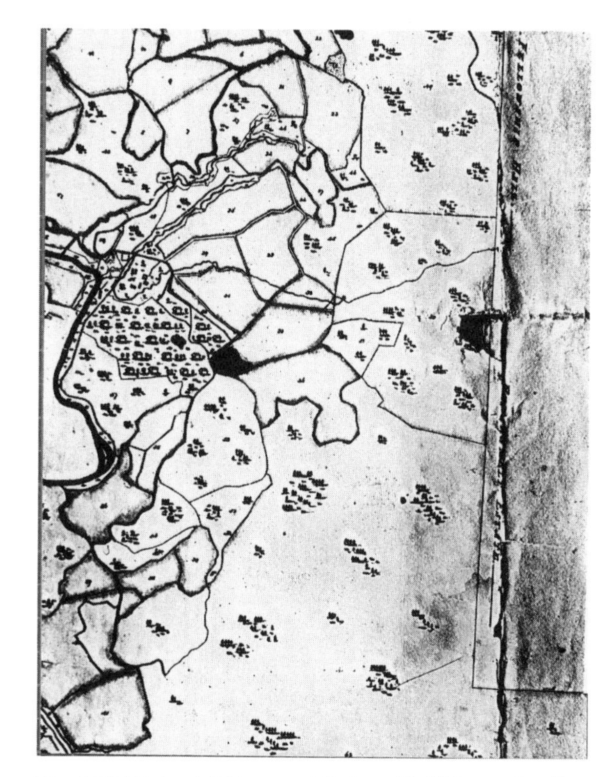

Fig. 6.10 Richard Wilson and Robert M. Wilson, detail from plan of Warwick Castle Estate, 1841 (Courtesy of the National Library of Jamaica)

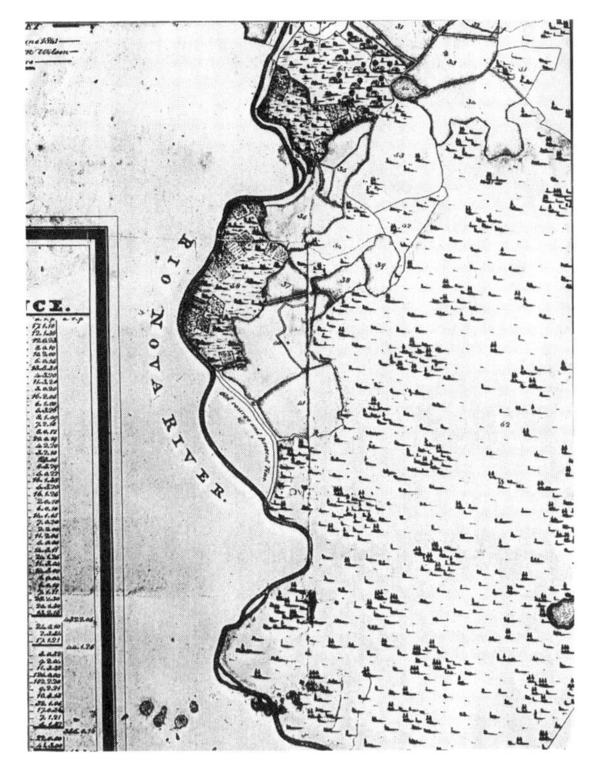

Fig. 6.11 Anonymous, detail from plan of Warwick Castle Estate, 1841 (Courtesy of the National Library of Jamaica)

just as the layout of the houses is symbolic. The number of houses shown, whether eighteen or fifteen, is too few to have sheltered the 271 slaves living on the estate in 1832 and must also be regarded as no more than symbolic.[26] The village area, numbered 63 on the plan, contained 33 acres. McGeachy's 1841 plan of the apparently abortive "new township" on Warwick Castle comprised seventy-five units, each one-quarter of an acre, or a total of only 19 acres.

The example of Warwick Castle suggests some common deficiencies in the representation of workers' housing on Jamaican plantation maps. Most obvious is that the number of houses depicted tends to be considerably fewer than the actual number. But basic patterns of layout are probably represented accurately, at least in a symbolic fashion. The two versions of the Warwick Castle plan must both be regarded as showing irregular, scattered layouts, not geometric patterns. Where houses are shown set out in regular lines or "ranges," it is fair to conclude that these patterns existed in fact rather than that the houses were scattered or clustered. This might seem a simple distinction, but it is important to an understanding of the extent of planter intervention in workers' housing and hence in the family and household organization of the plantation community. It is a question for which Jamaican plantation maps seem capable of supplying relatively reliable answers.

In the sample of 282 plantation maps, defined above, some 25 percent show the "village" area as a blank and so give no clues as to the layout of workers' housing. This proportion was particularly high before about 1770 and, to a lesser extent, after emancipation when some village sites were in fact deserted. Of those plans that do represent individual houses, the layout is irregular or dispersed on 78 percent of plantations and regular or linear on 22 percent. Of the latter, houses are laid out in straight lines in 78 percent of cases and in curvilinear, but geometric, patterns on 22 percent of plantations.

More interesting are the trends over time. The proportion of plans showing regular, linear layouts is low until 1790, reaches a peak around 1810, and declines after emancipation. The overall trend is smooth, and the peak period of geometric layouts suggests a relationship with the interventionist policies adopted by planters in many areas of slave life after 1790, sometimes referred to as an "amelioration" of conditions.[27] It relates also to the ideas of "improvement" common in British agriculture and the social engineering that went with industrialization. Similar tendencies in plantation layout occurred in Cuba and the French Antilles.[28]

In Jamaica, the proportion of geometric, regimented housing layouts is higher on the industrialized sugar estates and coffee plantations than on pens, with the highest ratio (40 percent) occurring on coffee plantations mapped between 1790 and 1810. Most of these were newly established plantations, and so could be laid out according to the planters' notions of spatial order and social control, whereas long-settled plantations were less likely to be transformed. But the proportion of symmetrical, linear layouts appearing on plans of sugar estates is almost as great as that on coffee plantations around 1800, and some planters certainly did reorganize their slave villages.

The general tendencies in the layout of workers' housing revealed by plantation maps, then, appear consistent and encourage faith in their reliability. This is not to say that individual plans necessarily always provided accurate data in terms of overall layout, but a sample as large as that analyzed here does seem quite robust. Data on numbers of houses are much less reliable. Some 64 percent of the plans show fewer than twenty houses, certainly a gross undercount, but the numbers shown vary little with layout.[29]

It is important to note, in concluding this discussion of workers' housing, that Jamaican slaves and their descendants lived almost always in detached houses or huts rather than in barracks or ranges divided into apartments. The universality of detached or semidetached housing in Jamaica distinguished the experience of the slaves from that typical of plantation communities in many parts of tropical America. Indentured laborers who came to Jamaica after emancipation, however, were commonly housed in barracks, and this pattern persisted into the twentieth century. Here the map evidence is weak. But the universality of detached housing during slavery had important implications for the development of Afro-Jamaican society, and the plans have much to offer in their detailed and sometimes precise information about the layout and architecture of plantation villages.[30]

Gardens and Grounds

The production of food for consumption on Jamaican plantations took place in gardens attached to great houses and workers' houses, in provision grounds located within the plantation boundaries, and in separate units of land called mountains because they were commonly situated in hilly backlands. The internal organization of these gardens, grounds, and mountains was not often surveyed in detail, but a few plans do provide unusually full information and deserve separate treatment. Great house kitchen gardens and ornamental gardens are included on plans relatively frequently, but the boundaries of gardens surrounding workers' houses are rarely distinguished. As to provision grounds, James Simpson, attorney for several absentee-owned estates in the eastern end of Jamaica from 1804 to 1828, stated in 1832 that "[i]t is not customary in Jamaica to make any survey of the land cultivated by the negroes, and they generally cultivate it in a straggling way, here and there where they find the best soil, if they had land enough to go upon, they cultivate that which is most easily cultivated and most productive, so that it is impossible to form a judgment of the extent of it in the aggregate."[31] This argument applied equally to provision grounds and separated mountains.

Great house gardens, whether kitchen or ornamental, followed the laws of order underlying the formal geometrical layout of European gardens, which predominated during the eighteenth century and were associated with dictatorial political systems.[32] The history of great house gardens has been little studied, and it is unknown whether garden designers practiced in Jamaica in the period 1700–1900.[33] Most plans of ornamental gardens on plantations show a simple square, crossed by diagonal paths, with a circle or fountain at the center. Such a layout may often have been no more than a conventional symbol, but it does suggest the basic principles.

The slaves' house gardens were generally fenced off, and only these boundaries appear, at best, on plantation maps. Examples of such plans, particularly the work of Pechon and the postemancipation surveyors, have been mentioned, but none provides any detail of the layout of the gardens themselves. Plans showing houses in perspective more often suggest the dominance of trees and shrubs in the village landscape, but only in a symbolic fashion. Houses are rarely set within gardens, and village gardens sometimes flow into neighboring areas of woodland, ruinate, and provision grounds.

Provision grounds, on which slaves were expected to produce their own food, existed on most Jamaican plantations after 1780. Only where almost all of the land within an estate was suited to the cultivation of sugar and backlands were not accessible did planters chose to purchase food or produce provision crops by supervised gang labor.[34] After emancipation, laborers who remained resident in plantation villages were also generally renters of provision grounds, though the planters continued to be disinclined to measure out precise areas, and the shifting character of land-use persisted.[35]

A distinction was sometimes made between provision grounds and home grounds. William Taylor, an estate attorney, stated in 1832 that the home ground was placed near the village and used to produce vegetables and yams, while the provision ground was used for sweet potatoes, corn, cocos, yams, plantains, and cassava.[36] Another attorney, William Miller, whose interests were on the north coast, stated that most plantations had "what they call dinner-time ground, that is ground that during the two hours' dinner-time, the negro can go to and return back to his work in the afternoon."[37] This area, said Miller, was located up to a mile from the slave village. The clearest example of this arrangement appears on Morris and Cunningham's 1830 plan of Retreat Pen in St. Ann. Retreat covered 1,040 acres and had a slave population of 235 in 1832.[38] There were as many as twelve units of "wood, ruinate and Negro grounds" on the fringes of the pen, with a total area of 719 acres, but there were 15 acres of "Great house provision grounds" 440 yards from the great house, 9 acres of "shell blow grounds" 750 yards from the village, and 2 acres of "provisions" 880 yards away. No doubt the "shell blow grounds" served the function of Miller's "dinner-time ground," but Morris and Cunningham provided no detail.

Plans purporting to show the subdivision of provision grounds occasionally suggest that they were laid out geometrically. John Rome's 1762 plan of Sandy Gully Pen in Clarendon (Vere) shows such an arrangement, though his boundary lines are improbably mechanical. Rome indicates no village site, so the "negro ground" may also have contained the houses. Vere was the prime area dependent on estate-produced food, and rigid control of grounds was more likely there than elsewhere. More realistic is a plan of about 1800 for Gayle Estate in St. Mary, showing clearings of irregular shape and indicating the range of crops cultivated.[39] All of these plans must be treated as providing little more than symbolic representations.

Where the land within a plantation's boundaries was all suited to the cultivation of export crops, planters frequently acquired blocks of land in surrounding uplands and required their slaves to use these "mountains" as provision grounds. Most mountains were less than 10 miles distant from the estates, but in some cases they were much farther away. Slaves were able to work these lands only on the days when they were not forced to perform estate labor, principally Sundays and occasional Saturdays, and necessarily planted a range of crops different than that cultivated on grounds close to the village.

The system of separated provision grounds was most common in the western parishes of Westmoreland, Hanover, St. James, and Trelawny, where the distinction between coastal plains and interior uplands was most marked. Carlton Estate, owned by the absentee Sir John Gordon, bounded the southeastern edge of Success Mountain in St. James. Carlton leased a small area of Success Mountain for cane land, but the 80 acres it owned on the northern boundary of Success Mountain was "run off and divided into lotts for the use of Negro grounds" when purchased in 1811 (figure 6.12). Here the surveyors, Stevenson and Smith, were required to order the land into units, and they did so by laying a geometric pattern over the land, oriented to the cardinal directions. Two areas were reserved, "not being very good" or "old land . . . to run up into ruinate," but the remaining units were squares and rectangles of varying size. Thus the lots had an average size of just over an acre. In 1811 there were 138 slaves on Carlton, giving a mean area of slightly more than half an acre of provision grounds per slave. In 1817, when detailed data became available, 70 of the slaves were males and 55 females.[40] The only lot to cover more than 2 acres (lot 16) was used by Driver Ned, a creole aged 47 in 1811; Co [Coromantee] Charlott, an African aged 34; Sandy, who apparently died between 1811 and 1817; and

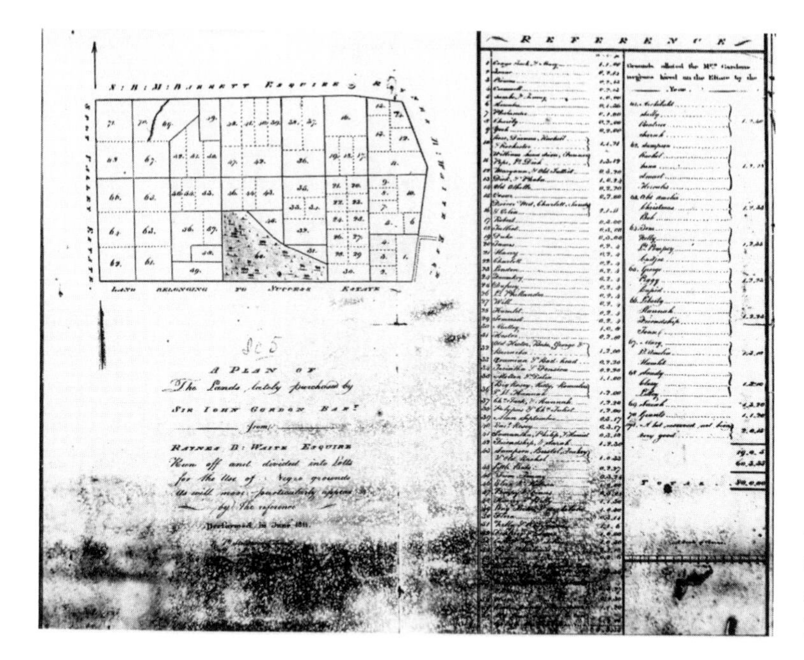

Fig. 6.12 Stevenson and Smith, *A Plan of the Lands Lately Purchased by Sir John Gordon, Bart.*, 1811 (Courtesy of the National Library of Jamaica)

Celia, Charlott's infant daughter. Lot 11 approached 2 acres and was used by William, the head driver, a creole aged 44; Fanney, a creole aged 30; and her two sons, Pope and Dick, aged 7 and 3 years. Fanney was the daughter of Juno, an African supposedly aged 74, who shared lot 52 with her son of 22 years, Venture. Lot 53 was used by Juno's other son, Cattleman Ned, aged 44, and Silvia, an African of the same age. Juno's daughter, Delia aged 24, shared lot 35 with Molton, an African man also aged 24. Most of the lots attributed to individuals were used by slaves lacking kin, and mature African males predominated in this group. Some of the lots listed for several slaves were used by family groups. Lot 36, for example, was used by Big Rosey, an African aged 42, and her three children aged 10, 5, and 1. Lots 12, 33, 41, 54, 56, and 57 were also used by mother-children families. The records do not reveal paternity, but the two drivers at least appear to have occupied units for their nuclear families. Thus Stevenson and Smith's plan of Carlton provision grounds can be made to yield rare and valuable information about the organization of slave households and the planter's perception of kinship links, when related to other documentary sources.[41]

Conclusion

Map evidence has rarely been given an equal place with other types of historical source materials. Historians have tended to emphasize the imprecision of much early cartographic work and to use it only in a strictly illustrative manner. Systematic analysis of map evidence remains uncommon, and the individual case is often taken to be typical. But maps in fact conform, in large measure, to the same procedures of analysis as other historical source materials, though the techniques employed in extracting their information may sometimes have to be borrowed from geography and cartography.[42]

The analysis of Jamaican plantation maps presented in this essay has sought to demonstrate their strong points as well as their weaknesses. Above all, it is important to assess each map in terms of its purpose and the techniques underlying its construction. The aims of the surveyor or planmaker are of paramount importance, and maps cannot be expected to yield accurate answers to questions that did not concern surveyors. Thus the many empty plans of early Jamaican plantations, produced strictly to provide boundary data, cannot be exploited to any significant extent as sources of information about the internal life of the plantation. Even plans that purported to show the internal layout of plantations provide evidence of unequal accuracy for different features. These variations in treatment all reflect the purposes and prejudices of surveyors and planmakers, and of their planter employers.

Only after the question of purpose has been examined is it appropriate to consider the effects of technique on the accuracy of representation. An assessment can be made, however, only in terms of the needs and expectations of the user. Historians are unlikely to require geodetic or cadastral precision and can readily accept data with a much grosser level of accuracy. The major weakness of Jamaican plantation mapping was the failure to orient surveys to the true azimuth. The disjunction in plantation boundaries became very common over time and the cause of many problems of land-tenure arrangements. But these difficulties need not concern the historian interested primarily in the internal structure of plantations, and at this level the degree of accuracy achieved by most plans produced after about 1750 is well within the limits of tolerance. Absolute precision is not to be looked for, the width of a pencil or ink line being sufficient in itself to create "errors." More important is the need to criticize the gaps in the information provided by plans. As J. B. Harley has said with regard to estate maps, "their interpretation, despite an apparent graphic simplicity, may bristle with difficulty."[43] All maps involve generalization and conventionalization, and an assessment of accuracy depends on distinguishing the "real" from the conventional and symbolic.

The unique advantage of map evidence is that it places data within a precise spatial framework. Thus it has been possible to use the Jamaican plantation maps to provide answers to many questions about the internal spatial ordering of estates, plantations, and pens, and to throw light on their social and economic organization. Although selective, the plans are more comprehensive than contemporary landscape art. For instance, planmakers were often ready to provide data on the layout and architecture of plantation workers' housing, whereas landscape artists were reticent about introducing squalid villages into their scenes. At the same time, the evidence provided by planmakers should not be pushed too hard. Most represented workers' housing in only a conventional manner, making no attempt to plot individual houses accurately. Even conventionalized layouts and architectural symbols provide useful data, however, if interpreted within the context of a large sample and over a long period of time.

Jamaican estate plans yield most when used in conjunction with other types of historical sources. In some cases, other sources show the weaknesses of map evidence. For example, the exploitation of logwood fences appears in plantation accounts, while the plans show merely the existence of fences. Most can be extracted when a sequence of plans covering a long period is available, but such a sequence is rare. The small number of plantation maps for the years before 1750 is also a serious lack, but this is a difficulty shared with most other types of Jamaican historical source materials. Obviously, it is a great advantage to have available a large sample of plans, enabling the kind of systematic analysis presented here, and unfortunately it will not be possible to carry out the same kind of work for all West Indian plantation societies.

Notes

1. J. B. Harley, "The Evaluation of Early Maps: Towards a Methodology," *Imago Mundi* 22 (1968): 62–74; Harley, *Maps for the Local Historian: A Guide to the British Sources* (London, 1972). See also David Buisseret, ed., *From Sea Charts to Satellite Images: Interpreting North American History through Maps* (Chicago, 1990).

2. See Arthur H. Robinson and Barbara Bartz Petchenik, *The Nature of Maps* (Chicago, 1976); J. S. Keates, *Understanding Maps* (London, 1982).

3. Cf. J. H. Andrews, *Irish Maps* (Dublin, 1978).

4. For the adequacy of these measures see Joan Murphy, "Measures of Map Accuracy Assessment and Some Early Ulster Maps," *Irish Geography* 11 (1978): 88–101.

5. Cf. Harley, "Evaluation of Early Maps," 65.

6. For full quantitative analyses, see B. W. Higman, "The Spatial Economy of Jamaican Sugar Plantations: Cartographic Evidence from the Eighteenth and Nineteenth Centuries," *Journal of Historical Geography* 13 (1987): 17–39; Higman, "Jamaican Coffee Plantations, 1780–1860: A Cartographic Analysis," *Caribbean Geography* 2 (1986): 73–91; Higman, "The Internal Economy of Jamaican Pens, 1760–1890," *Social and Economic Studies* 38 (1989): 61–86. For additional illustrations and examples, see Higman, *Jamaica Surveyed: Plantation Maps and Plans of the Eighteenth and Nineteenth Centuries* (Kingston, 1988).

7. Noel Deerr, *The History of Sugar* (London, 1949–50), 1:198–99; Richard B. Sheridan, *Sugar and Slavery: An Economic History of the British West Indies, 1623–1775* (Barbados, 1974), 487–89.

8. Douglas Hall, *Free Jamaica, 1838–1865* (New Haven, 1959), 84, 95; R. W. Beachey, *The British West Indies Sugar Industry in the Late Nineteenth Century* (Oxford, 1957), 129.

9. Gisela Eisner, *Jamaica, 1830–1930* (Manchester, 1961), 238.

10. Ibid.

11. B. W. Higman, *Slave Population and Economy in Jamaica, 1807–1834* (Cambridge, 1976), 17.

12. Ibid., 14; Hall, *Free Jamaica,* 82; Deerr, *History of Sugar* 1:177; Eisner, *Jamaica,* 299; *Handbook of Jamaica,* 1882 and 1900. The sources do not always agree, and these numbers should be taken only as indicators of general trends.

13. *Jamaica House of Assembly Journals* 10 (1799): 438.

14. Higman, *Slave Population and Economy in Jamaica,* 224; *Jamaica House of Assembly Votes,* 1847, 373; *Handbook of Jamaica,* 1900, 411.

15. Higman, *Slave Population and Economy in Jamaica,* 25–26; Edward Long, *The History of Jamaica* (London, 1774), 1:380; *Handbook of Jamaica,* 1882, 491–93; 1900, 403; 1915, 442–50. The first detailed list in the *Handbook of Jamaica* includes four hundred "grazing pens having 100 heads of cattle" in 1913–14.

16. For discussion of the problems of measuring shape see Peter Haggett, *Locational Analysis in Human Geography* (London, 1965), 227; H. Kishimoto, *Cartometric Measurements* (Zurich, 1968); D. H. Maling, "How Long Is a Piece of String?" *Cartographic Journal* 5 (1968): 147–56; Jack P. Gibbs, ed., *Urban Research Methods* (Princeton, 1961), 101–3.

17. Higman, *Slave Population and Economy in Jamaica,* 51–52, 246. Average slope is derived from the number of grid/contour crossings, by Wentworth's formula (G. H. Dury, *Map Interpretation* [London, 1960], 176).

18. Robertson's map of 1804, and Accounts Produce, Liber 72, fol. 226, Jamaica Archives, Spanish Town; *Jamaica Almanack,* 1817 and 1837; *Handbook of Jamaica,* 1883, 387.

19. Accounts Produce, Liber 73, fol. 219; and Returns of Registrations of Slaves, 1832, Jamaica Archives, Spanish Town; *Handbook of Jamaica,* 1883, 382.

20. P. J. Laborie, *The Coffee Planter of Santo Domingo* (London, 1798), 13.

21. Accounts Produce, Liber 73, fol. 128; *Jamaica Almanack,* 1832; Returns of Registrations of Slaves, 1832; *Jamaica House of Assembly Votes,* 1847, 373.

22. J. Manderson, plan of Fullerswood and Beach Pens, 1864, National Library of Jamaica plans, St. Elizabeth B; Accounts Produce, Liber 73, fols. 30–31; *Jamaica Almanack,* 1832.

23. Journal of Benjamin Scott Moncrieffe, St. Ann, 1828–40 (Account Book of Union Pen), Jamaica Archives, Spanish Town; *Jamaica Almanack,* 1832.

24. Higman, *Jamaica Surveyed,* chap. 8.

25. Commons "Report from Select Committee on the Extinction of Slavery throughout the British Dominions," *Parliamentary Papers,* 1832, vol. 20, sessional paper 721, pp. 29, 169; Hope Masterton Waddell, *Twenty-nine Years in the West Indies and Central Africa* (London, 1863), 147.

26. Returns of Registrations of Slaves, 1832.

27. Cf. Edward Brathwaite, *The Development of Creole Society in Jamaica, 1770–1820* (Oxford, 1971), 234.

28. See Helen Rosenau, *Social Purpose in Architecture: Paris and London Compared, 1760–1800* (London, 1970); Francis D. Klingender, *Art and the Industrial Revolution* (Frogmore, England, 1972); Manuel Moreno Fraginals, *El ingenio: Complejo economico social cubano del azúcar* (Havana, 1978), 2:68–83; Gabriel Debien, *Les Esclaves aux Antilles françaises, XVIIe–XVIIIe siècles* (Basse-Terre, Guadeloupe, 1974), 219–34.

29. Cf. Horace Fairhurst, "The Surveys for the Sutherland Clearances, 1813–1820," *Scottish Studies* 8 (1964): 6.

30. Cf. Sidney W. Mintz and Richard Price, *An Anthropological Approach to the Afro-American Past: A Caribbean Perspective* (Philadelphia, 1976).

31. "Select Committee on the Extinction of Slavery," 385.

32. See Julia S. Berrall, *The Garden: An Illustrated History* (Harmondsworth, 1978); A. A. Tait, *The Landscape Garden in Scotland, 1735–1835* (Edinburgh, 1980); William Howard Adams, *The French Garden, 1500–1800* (New York, 1979).

33. K. E. Ingram, *Sources of Jamaica History, 1655–1838,* (Zug, 1976), 2:931–33; Douglas Hall, *Planters, Farmers, and Gardeners in Eighteenth Century Jamaica* (Mona, Jamaica, 1988).

34. Orlando Patterson, *The Sociology of Slavery* (London, 1967), 218; Higman, *Slave Population and Economy in Jamaica,* 122.

35. Hall, *Free Jamaica,* 174–75.

36. "Select Committee on the Extinction of Slavery," 33.

37. "Report from the Select Committee on Negro Apprenticeship in the Colonies," *Parliamentary Papers,* 1836, sessional paper 560, p. 339.

38. Morris and Cunningham, plan of Retreat Pen, 1830, National Library of Jamaica plans, St. Ann 21; Returns of Registrations of Slaves, 1832.

39. John Rome, plan of Sandy Gully Pen, 1762, National Library of Jamaica, Dawkins MS, vol. 7; plan of Gayle Estate, c. 1800, National Library of Jamaica plans, St. Mary 11.

40. Returns of Registration of Slaves, 1817, Liber 31, fol. 140. See also *Jamaica Almanack,* 1811 (138 slaves); Accounts Produce, Liber 73, fol. 63.

41. Cf. B. W. Higman, "The Slave Family and Household in the British West Indies, 1800–1834," *Journal of Interdisciplinary History* 6 (1975): 261–87.

42. Harley, "Evaluation of Early Maps," 73.

43. Harley, *Maps for the Local Historian,* 36.

CONCLUSION: THE INCIDENCE *and* SIGNIFICANCE *of* ESTATE MAPS

DAVID BUISSERET

Our odyssey from late sixteenth-century England to late nineteenth-century Jamaica is now complete. Along the way, we have found estate maps in a wide variety of countries, with very different incidences in time and space. Probably the most important element governing their appearance was the existence of large agricultural units of valuable land. For instance, the prosperous riverine plantations of South Carolina generated a wealth of estate plans, almost totally lacking from the hardscrabble farms of North Carolina. Similarly, when the *arpenteurs* eventually began work in France, it was on the prosperous farms of the northern plain, and not among the impoverished smallholders of the Cévennes.

Estate plans also seem to be connected with a lively market in land. They flourished in late sixteenth-century England, where the earlier depredations of Henry VIII led to wholesale and continuing transfers of land, but not in the rich river valleys of France and Spain, where on the whole the patterns of land tenure were much less volatile. It seems, too, that their incidence coincides to some degree, in Europe, with exposure to the ideas of what has become called the Agricultural Revolution. Here the contrast between England on the one hand and seventeenth-century France and Spain on the other is very striking.

Often, an inclination to take the land in hand, in order to satisfy the demands of the market, went along with a heightened map consciousness among landowners. Where this did not exist, as among the Junker of Prussia, no amount of other favorable factors could conjure estate maps. These did not always emerge even when such consciousness existed. Sometimes the task of the landowner had already been assumed by some local authority, as in the case of Venetian Terra Ferma or the Netherlands, and sometimes, as in much of Germany, landowners were so accustomed to a different form of delineation (the *Landtafel*) that they did not readily take to the estate map. In the New World the relationship between estate maps and production for massive, often overseas markets seems established by the examples of Jamaica, St. Domingue, and South Carolina.

Surveyors need a modicum of civil order in order to go about their business, and this was singularly lacking for long periods in sixteenth- and seventeenth-century France and in seventeenth-century Germany. We have no assessment of the effect of the Civil War upon estate mapping in England, but it would be surprising if there were not a marked reduction, or even cessation, of activity in the 1640s and 1650s.

Finally, of course, estate maps could not emerge or thrive in regions that already had large-scale topographical coverage. Here the classic example is Ireland, whose numerous surveyors virtually ceased to operate after the Ordnance Survey finished its work in the 1830s and 1840s. Well-developed systems of cadastral state survey, like those described by Kain and Baigent for many of the Scandinavian countries,[1] could also preempt even the emergence of estate maps.

Estate maps could take widely differing forms. The rather misleading English term had its counterparts in German, *Flurkarte,* French, *plan parcellaire seigneurial,* and Dutch, *domanial kaart,* indicating that the type was well recognized. Almost all were manuscript, usually on paper, and many were colored. Sometimes, as in Jamaica and South Carolina, many of the surviving examples are rough copies; this seems much less common in England. It is also true that in Jamaica and South Carolina the celebratory element is often lacking; these maps were often strictly working copies.

Estate plans came in a wide variety of styles, with an infinite variety of cartouches and forms of lettering. At first, they rarely gave any indication of relief, but as time went by there were precocious mapmakers, like Joseph Purcell of South Carolina, Jean Bonnet Pechon of Jamaica, and Bernard Scalé in Ireland, who began introducing various forms of hachuring. Most of the maps stand individually, but in England and Ireland, particularly for the estates of great lords, they were sometimes bound into atlases of uniform format; this also occurred occasionally in Jamaica but never, as far as can be told, in South Carolina.

It is very difficult to offer any figures for the numbers of estate maps produced over the centuries in the different countries. For Jamaica, where the collection is relatively concentrated, Higman has been able to produce convincing figures, but this information would be very hard to collect for other areas, where the sources are much more scattered. Perhaps the best that we can do is to work from the numbers of surveyors known to have been active. For Great Britain and Ireland, between 1550 and 1850, Peter Eden enumerates rather more than seven thousand of them.[2] In chapter 5 Higman has explained that the number working in Jamaica between 1700 and 1900 was about three hundred; this was about the same as the number working in South Carolina between 1670 and 1775. Higman is able to offer some comparison between Jamaica and Scotland, but it seems impossible to extend this calculation to the British Isles as a whole, because we do not have figures for the number of surveyors working in any given year. It would be possible, though, to break down Eden's figures by decade, and so to see what sort of pattern emerged. The *arpenteurs* and *landmeter* of continental Europe remain even more mysterious, for we do not even have good general lists of them.

It is possible to discover a little more about the social standing of the anglophone surveyors. In Ireland it seems to have been very varied, ranging from fashionable Dublin surveyors like John Rocque and Bernard Scalé, to backcountry squires or worse.[3] In England there was probably a comparable range, to judge by the soundings carried out by Sarah Bendall.[4] In Jamaica, too, there was a "considerable range of individual variation" in the fortunes of surveyors.[5] There may be a contrast here with the American mainland. In Virginia, Sarah Hughes notes, many surveyors became leading members of society, partly because of the opportunities that came to them in the course of their work;[6] they were often the first Europeans to come across potentially fertile land.[7] The same seems to be true, to a lesser degree, in South Carolina. The fourteen prolific surveyors noted in chapter 4 were nearly all prominent in colonial society; some drew maps of large areas, others became substantial landowners, and others again were connected to powerful families.

At least one of the South Carolina surveyors was also a highly competent engineer: Charles Blacker Vignoles, who was surveying South Carolina estates in the early nineteenth century, later became known internationally for his engineering skills.[8] By 1800 this would have been an unlikely combination in the British Isles, for as Andrews explains, engineering was then "moving rapidly ahead of land surveying in technical complexity, in educational standards, and in the cost and precision of the apparatus required; also in prestige and profitability."[9] The age of railroads and canals required surveys of very great accuracy, particularly in the depiction of relief, which was not an area in which most estate surveyors excelled.

The training of surveyors in all the countries that we have been considering relied essentially upon an apprenticeship system, followed on occasion by an examination. The vast and often heavily wooded areas of the Americas encouraged general reliance upon the relatively crude system of the compass and chain, whereas the surveyors of England and Ireland, dealing with smaller and more valuable plots, often in open ground, could use plane tabling and triangulation; the coming of newly accurate theodolites in the 1730s accentuated this trend toward greater accuracy.[10]

Curiously, it seems impossible to establish any relationship between surveyors on different sides of the Atlantic, or even of the Irish Sea and the Strait of Florida; it is as if each "school" of surveyors developed independently, using the apprenticeship system to pass on skills. For instance, the only connection that we can establish between the West Indies and the Carolinas is John Love's *Geodesia* of 1688; as we have seen, he had worked in both Jamaica and North Carolina, and his book had two American editions as well as numerous British ones.

Sarah Bendall and B. W. Higman have made careful assessments of the accuracy of the plans that they have studied, and Fritz Bönisch has done the same for many German plans.[11] In this volume, readers may judge from those plates that have comparative modern maps just how remarkably closely the estate surveyors often approximated reality. Their chosen realm—a world of fields, woods, streams, farm buildings, and dwellings—is almost unrecoverable by other, verbal means, and the maps that they left are our most precious resource for imaginatively reconstructing a world that we have lost.

Notes

1. Roger Kain and Elizabeth Baigent, *The Cadastral Map in the Service of the State: A History of Property Mapping* (Chicago, 1992), 49–119.

2. Peter Eden, ed., *Dictionary of Land Surveyors and Local Cartographers of Great Britain and Ireland, 1550–1850* (Folkestone, 1975–76).

3. J. H. Andrews, *Plantation Acres: An Historical Study of the Irish Land Surveyor and His Maps* (Belfast, 1985), 269.

4. Sarah Bendall, *Maps, Land, and Society: A History, with a Carto-Bibliography of Cambridgeshire Estate Maps, c. 1600–1836* (Cambridge, 1992), 104–14.

5. B. W. Higman, *Jamaica Surveyed: Plantation Maps and Plans of the Eighteenth and Nineteenth Centuries* (Kingston, 1988), 39.

6. Sarah S. Hughes, *Surveyors and Statesmen: Land Measuring in Colonial Virginia* (Richmond, 1979), 4–5.

7. A point noted by F. M. L. Thompson, *Chartered Surveyors: The Growth of a Profession* (London, 1968), 44.

8. K. H. Vignoles, "Charles Blacker Vignoles in South Carolina and Florida, 1817–1823," *South Carolina Historical Magazine* 85 (1984): 83–107.

9. Andrews, *Plantation Acres,* 254.

10. Thompson, *Chartered Surveyors,* 27.

11. Bendall, *Maps, Land, and Society,* 50–76; Higman, *Jamaica Surveyed,* 78–79; Fritz Bönisch, "The Geometrical Accuracy of Sixteenth- and Seventeenth-Century Topographical Drawings," *Imago Mundi* 21 (1967): 62–69.

BIBLIOGRAPHY

Adams, Ian H. "The Land Surveyor and His Influence on the Scottish Rural Landscape." *Scottish Geographical Magazine* 84 (1968): 248–55.

———. *The Mapping of a Scottish Estate.* Edinburgh, 1971.

Adams, William Howard. *The French Garden, 1500–1800.* New York, 1979.

Agas, Ralph. *A Preparative to Platting of Landes and Tenements for Surveigh.* London, 1596.

Alfrey, Nicholas, and Stephen Daniels, eds. *Mapping the Landscape: Essays on Art and Cartography.* Nottingham, 1990.

Andrews, J. H. "The French School of Dublin Land Surveyors." *Irish Geography* 5 (1967): 275–92.

———. *Irish Maps.* Dublin, 1978.

———. *A Paper Landscape: The Ordnance Survey in Nineteenth-Century Ireland.* Oxford, 1975.

———. *Plantation Acres: An Historical Study of the Irish Land Surveyor and His Maps.* Belfast, 1985.

Angelini, Gregorio, ed. *Il disegno del territorio: Istituzioni e cartografia in Basilicata, 1500–1800.* Exhibit catalog. Bari, 1988.

Archives de France. *Espace français: Vision et aménagement, XVIe–XIXe siècles.* Exhibit catalog. Paris, 1987.

———. *Voyage aux Iles d'Amérique.* Exhibit catalog. Paris, 1992.

Aylmer, G. E. "The Economics and Finances of the Colleges and University, c. 1530–1640." In *The History of the University of Oxford,* vol. 3, *The Collegiate University,* ed. James McConica. Oxford, 1986.

Aymans, Gerhard. *Landkarten als Geschichtsquellen.* Exhibit catalog. Cologne, 1985.

Barbier, Frédéric, ed. *La Carte manuscrite et imprimée du XVIe au XIXe siècle.* New York, 1983.

Baricchi, Walter, ed. *Le mappe rurali del territorio di Reggio Emilia: Agricoltora e paisaggio tra XVI e XIX secolo.* Casalecchio di Reno Bologna, 1985.

Barnett, Lloyd G. *The Constitutional Law of Jamaica.* Oxford, 1977.

Bartram, Alan. *Tombstone Lettering in the British Isles.* London, 1978.

Beachey, R. W. *The British West Indies Sugar Industry in the Late Nineteenth Century.* Oxford, 1957.

Beastall, T. W. *A North Country Estate: The Lumleys and Saundersons as Landowners, 1600–1900.* London, 1975.

Becker, Robert H., ed. *Designs on the Land: Diseños of California Ranches.* San Francisco, 1969.

Behr, Hans-Joachim, et al. *Geschichte in Karten: Historische Ansichten aus den Rheinland und Westfalen.* Exhibit catalog. Düsseldorf, 1985.

Bendall, Sarah. *Maps, Land, and Society: A History, with a Carto-Bibliography of Cambridgeshire Estate Maps, c. 1600–1836* Cambridge, 1992.

———, ed., *Peter Eden's Dictionary of Land Surveyors of Great Britain and Ireland, c. 1540–1850.* 2d ed. Forthcoming.

Benes, Peter. *New England Prospect: A Loan Exhibition of Maps at the Currier Gallery of Art.* Boston, c. 1981.

Benzing, Josef. *Jacob Köbel zu Oppenheim, 1494–1533.* Wiesbaden, 1952.

Beresford, M. W. *History on the Ground.* London, 1957.

———. "Ridge and Furrow and the Open Fields." *Economic History Review,* 2d ser., 1 (1948–49): 34–45.

Berrall, Julia S. *The Garden: An Illustrated History.* Harmondsworth, 1978.

Betts, Edwin Morris. *Thomas Jefferson's Farm Book.* Princeton, 1953.

Bloch, Marc. "Les Plans Parcellaires." *Annales d'Histoire Economique et Sociale* 1 (1929): 60–70, 390–98.

Bönisch, Fritz. "The Geometrical Accuracy of Sixteenth- and Seventeenth-Century Topographical Surveys." *Imago Mundi* 21 (1967): 62–69.

Bönisch, Fritz, Hans Brichzin, Klaus Schillinger, and Werner Stams. *Kursäschische Kartographie bis zum Dreissigjährigem Krieg.* Berlin, 1990.

Brathwaite, Edward. *The Development of Creole Society in Jamaica, 1770–1820.* Oxford, 1971.

Brethauer, Karl. "Johannes Krabbe Mundensis." *Braunschweigisches Jahrbuch* 55 (1974): 72–89.

Bridenbaugh, Carl. *Myths and Realities: Societies of the Colonial South.* Baton Rouge, 1952.

Buisseret, David, ed. *From Sea Charts to Satellite Images: Interpreting North American History through Maps.* Chicago, 1990.

———. *Mapping the French Empire in North America.* Exhibit catalog. Chicago, 1991.

———. *Monarchs, Ministers, and Maps: The Emergence of Cartography as a Tool of Government in Early Modern Europe.* Chicago, 1992.

———. *Rural Images: The Estate Plan in the Old and New Worlds.* Exhibit catalog. Chicago, 1988.

Buisseret, David, with Jack Tyndale-Biscoe. *Historic Jamaica from the Air.* Barbados, 1968.

Butlin, Robin. "Small-Scale Urban and Industrial Development in North-East Cambridgeshire in the Nineteenth and Early Twentieth Centuries." In *The Transformation of Rural Society, Economy, and Landscape: Papers from the 1987 Meeting of the Permanent European Conference for the Study of the Rural Landscape,* ed. Ulf Sporrong, 217–26. Stockholm, 1990.

Chadwick, S. J. *Dewsbury Moot Hall: Account Rolls and Court Roll.* Leeds, 1911.

Clausner, Marlin D. *Rural Santo Domingo: Settled, Unsettled, and Resettled.* Philadelphia, 1973.

Clemenson, H. A. *English Country Houses and Landed Estates.* London, 1982.

Conzen, Michael P. "The County Landownership Map in America: Its Commercial Development and Social Transformation, 1814–1939." *Imago Mundi* 36 (1984): 9–31.

———, ed. *The Making of the American Landscape.* Boston, 1990.

Corbitt, Duvon C. "Mercedes and Realengos: A Survey of the Public Land System in Cuba." *Hispanic American Historical Review* 19 (1939): 262–85.

Cosgrove, Denis. *The Palladian Landscape: Geographical Change and Its Cultural Representations in Sixteenth-Century Italy.* Leicester, 1993.

Cosgrove, Denis, and Geoff Pett, eds. *Water, Engineering,*

and Landscape: Water Control and Landscape Transformation in the Modern Period. London, 1990.

Cox, Nicholas. *Bridging the Gap: A History of the Corporation of the Sons of the Clergy over 300 Years, 1655–1978.* Oxford, 1978.

Cumming, William P. *British Maps of Colonial America.* Chicago, 1974.

Daelens, F. H. van der Haegen, and Eduard van Ermen, eds. *Oude kaarten en plattegronde: Brinnen voor de historische geografie van dezuidelijke Nederlanden, 16de–18de eeuw.* Brussels, 1986.

Dale, P. F. *Cadastral Surveys within the Commonwealth.* London, 1976.

Danforth, Susan, ed. *The Land of Norumbega: Maine in the Age of Exploration and Settlement.* Portland, 1988.

Danfrie, Philippe. *Declaration de l'usage du graphometre.* Paris, 1597.

Davies, O. R. F. "The Wealth and Influence of John Holles, Duke of Newcastle, 1694–1711." *Renaissance and Modern Studies* 9 (1965): 22–46.

Debien, Gabriel. *Les Esclaves aux Antilles françaises, XVIIe–XVIIIe siècles.* Basse-Terre, Guadeloupe, 1974.

———. *La Sucrerie Galbaud du Fort, 1690–1802.* N.p., 1941.

Deerr, Noel. *The History of Sugar.* London, 1949–50.

Desreumaux, Roger. "Sources Géographiques concernant la France conservées aux Archives Capitulaires de Tournai." *Horae Tornacenses* (1971): 275–91.

Dilich, Wilhelm. *Wilhelm Dilichs Landtafeln Hessischer Ämter zwischen Rhein und Weser.* Ed. Edmund Stengel. Marburg, 1972.

Donkersloot–de Vrij, Ypkje Marijke. *Topografische kaarten van Nederland vóór 1750.* Groningen, 1981.

———. *De Vechstreek: Oude kaarten en de geschiedenis van het landschap.* Groningen, 1985.

Dunbabin, J. P. D. "College Estates and Wealth, 1660–1815." In *The History of the University of Oxford,* vol. 5, *The Eighteenth Century,* ed. L. S. Sutherland and L. G. Mitchell. Oxford, 1986.

Dury, G. H. *Map Interpretation.* London, 1960.

Dymond, David. *Israel Amyce's Map of Melford Manor, 1580.* Long Melford, England, 1987.

Eckhardt, Wilhelm Alfred. "Wilhelm Dilichs Zehnkarte von Niederzwehren." *Zeitschrift des Vereins für Hessische Geschichte und Landeskunde* 72 (1961): 99–121.

Eden, Peter, ed. *Dictionary of Land Surveyors and Local Cartographers of Great Britain and Ireland, 1550–1850.* Folkestone, 1975–76.

Edwards, A. C., and K. C. Newton. *The Walkers of Henningfield.* London, 1984.

Eisner, Gisela. *Jamaica, 1830–1930.* Manchester, 1961.

Emlen, Robert P. *Shaker Village Views.* Hanover, N.H., 1987.

Emmerich, Werner. "Flurpläne aus der Zeit des sächsischen Kurfürsten Friedrich August I im Leipziger Stadtarchiv." *Jahrbuch für die Geschichte Mittel- und Ostdeutschlands* 11 (1962): 111–35.

Emmison, Frederick G. "Estate Maps and Surveys." *History* 48 (1963): 34–37.

Engel, Werner. "Joist Moers im dienste des Landgrafen Moritz von Hessen." *Hessisches Jahrbuch für Landesgeschichte* 32 (1982): 165–73.

Ernst, Joseph W. *With Compass and Chain: Federal Land Surveyors in the Old Northwest, 1785–1816.* New York, 1979.

Evans, I. M., and H. Lawrence. *Christopher Saxton: Elizabethan Map-Maker.* Wakefield, England, 1979.

Faini, Sandra, and Luca Majoli. *La Romagna nella cartografia a stampa dal cinquecento all'ottocento.* Rimini, 1992.

Fairhurst, Horace. "The Surveys for the Sutherland Clearances, 1813–1820." *Scottish Studies* 8 (1964): 1–18.

Flint, Abel. *A System of Geometry and Trigonometry: Together with a Treatise on Surveying.* Hartford, 1813.

Fockema Andreae, Sybrandus Johannes, and B. van t'Hoff. *Geschiedenis der kartografie van Nederland van den Romeinschen tijd tot het midden der 19de eeuw*. The Hague, 1947.

Folkerts, Menso, ed. *Mass, Zahl, und Gewicht: Mathematik als Schlüssel zu Weltverständnis und Weltbeherrschung.* Exhibit catalog. Weinheim, 1989.

Folkingham, William. *Feudigraphia: The Synopsis or Epitome of Surveying Methodized.* London, 1610.

Fougères, M. "Les Plans cadastraux de l'ancien régime." *Mélanges d'Histoire Sociale, Annales d'Histoire Sociale* 3 (1945): 54–69.

Fowkes, D. V., and G. R. Potter, eds. *William Senior's Survey of the Estates of the First and Second Earls of Devonshire, c. 1600–1628.* Chesterfield, England, 1988.

Gagel, Ernst. *Pfinzing: Der Kartograph der Reichsstadt Nürnberg, 1554–1599.* Hersbruck, 1957.

Geggus, David Patrick. *Slavery, War, and Revolution: The British Occupation of Saint Domingue, 1793–1798.* Oxford, 1982.

Gibbs, Jack P., ed. *Urban Research Methods.* Princeton, 1961.

Gittenberger, Franz, and Helmut Weiss. *Zeeland in oude kaarten.* Knokke, Belgium, 1983.

Granger, Mary, ed. *Savannah River Plantations.* Savannah, 1947.

Grant, V. B. *Jamaican Land Law.* Kingston, 1957.

Gray, H. L. *English Field Systems.* Cambridge, Mass., 1915.

Greene, Jack. "Colonial South Carolina and the Caribbean Connection." *South Carolina Historical Magazine* 88 (1987): 192–210.

Grenacher, Franz. "Daniel Meyer: Ein unbekannter schweizerischer Kartograph und der Kataster seiner Zeit." *Geographica Helvetica* 15 (1960): 8–16.

Grosjean, Georges, and Rudolf Kinauer. *Kartenkunst und Kartentecknik von Altertum bis zum Barock.* Bern and Stuttgart, 1970.

Gunter, Edmund. *Use of the Sector, Crosse-Staffe, and Other Instruments.* London, 1624. Reprint. Amsterdam, 1971.

Haggett, Peter. *Locational Analysis in Human Geography.* London, 1965.

Hall, Douglas. *Free Jamaica, 1838–1865.* New Haven, 1959.

———. *Planters, Farmers, and Gardeners in Eighteenth Century Jamaica.* Mona, Jamaica, 1988.

Harley, J. B. "The Evaluation of Early Maps: Towards a Methodology." *Imago Mundi* 22 (1968): 62–74.

———. *Maps for the Local Historian: A Guide to the British Sources.* London, 1972.

Harms, Hans. *Themen alter Karten.* Exhibit catalog. Oldenburg, 1979.

Harris, John. *The Artist and the Country House: A History of Country House and Garden View Painting in Britain, 1540–1870.* London, 1979.

Harris, Richard Colebrook, and John Warkentin. *Canada before Confederation: A Study in Historical Geography.* New York, 1974.

Hartog, Johannes. *Curaçao: From Colonial Dependence to Autonomy.* Aruba, 1968.

Harvey, P. D. A. *The History of Topographical Maps: Symbols, Pictures, and Surveys.* London, 1980.

———. *Manorial Records.* London, 1984.

———. *Maps in Tudor England.* Chicago, 1993.

———. "The Portsmouth Map of 1545 and the Introduction of Scale-Maps into England." In *Hampshire Studies,* ed. J. Webb, N. Yates, and S. Peacock. Portsmouth, England, 1981.

Harvey, P. D. A., and H. Thorpe. *The Printed Maps of Warwickshire, 1576–1900.* Warwick, 1959.

Hellwig, Fritz. "Tyberiade und Augenschein: Zur forensischen Kartographie im 16. Jahrhundert." In *Festschrift für Bodo Borner.* In preparation.

Hellwig, Fritz, Wolfgang Reiniger, and Klaus Stopp. *Landkarten der Pfalz am Rhein, 1513–1803.* Exhibit catalog. Bad Kreuznach, 1984.

Heslinga, M. W., A. P. de Klerk, H. Schmal, T. Stol, and A. J. Therkow. *Nedeland in kaarten.* Ede, 1985.

Hessischen Staatsarchiv. *Hessen im Bild alter Landkarten.* Marburg, 1988.

Higman, B. W. "Jamaican Coffee Plantations, 1780–1860: A Cartographic Analysis." *Caribbean Geography* 2 (1986): 73–91.

———. *Jamaica Surveyed: Plantation Maps and Plans of the Eighteenth and Nineteenth Centuries.* Kingston, 1988.

———. "The Slave Family and Household in the British West Indies, 1800–1834." *Journal of Interdisciplinary History* 6 (1975): 261–87.

———. *Slave Population and Economy in Jamaica, 1807–1834.* Cambridge, 1976.

———. "The Spatial Economy of Jamaican Sugar Plantations: Cartographic Evidence from the Eighteenth and Nineteenth Centuries." *Journal of Historical Geography* 13 (1987): 17–39.

Hilliard, S. B. "Antebellum Tidewater Rice Culture in South Carolina and Georgia." In *European Settlement and Development in North America,* ed. James R. Gibson. Toronto, 1978.

———. *Atlas of Antebellum Southern Agriculture.* Baton Rouge, 1984.

———. "Plantations and the Molding of the Southern Landscape." In *The Making of the American Landscape,* ed. Michael P. Conzen. Boston, 1990.

———. "Site Characteristics and Spatial Stability of the Louisiana Sugarcane Industry." *Agricultural History* 53 (1979): 254–69.

Hindle, Paul. *Maps for Local History.* London, 1988.

Historisches Archiv der Stadt Köln. *Alte handgezeichnete Kölner Karten.* Exhibit catalog. Cologne, 1977.

Hodson, D. *Maps of Portsmouth before 1801.* Portsmouth, England, 1978.

Höhn, Alfred. *Franken im Bild alter Karten.* Exhibit catalog. Würzburg, 1986.

Hopkins, Daniel. "An Extraordinary Eighteenth-Century Map of the Danish Sugar-Plantation Island St. Croix." *Imago Mundi* 41 (1989): 44–58.

———. "Jens Michelsen Beck's Map of a Danish West Indian Sugar-Plantation Island." *Terrae Incognitae* 25 (1993): 99–114.

Hughes, Sarah S. *Surveyors and Statesmen: Land Measuring in Colonial Virginia.* Richmond, 1979.

Hulsius, Levinus. *Erster Tractat der mechanischen Instrumenten.* Frankfurt, 1603.

Hyde, R. "Thomas Hornor: Pictorial Land Surveyor." *Imago Mundi* 29 (1977): 23–34.

Ingram, K. E. *Sources of Jamaica History, 1655–1838.* Zug, 1976.

Jacob, G. *The Complete Court-Keeper; or, Land-Steward's Assistant.* 8th ed. London, 1819.

Jäger, Eckhardt. *Lüneburger Beiträge zur Vedutenforschung.* Exhibit catalog. Lüneburg, 1983.

———. *Prussia-Karten, 1542–1810.* Weissenhorn, 1982.

Jenkins, Philip. "Cambridgeshire and the Gentry: The Origins of a Myth." *Journal of Local and Regional Studies* 4 (1984): 1–17.

Kahlfuss, Hans-Jürgen. *Landesaufnahme und Flurvermessung in den Herzogtümern Schleswig, Holstein, und Lauenburg vor 1864.* Neumünster, 1969.

Kain, Roger, and Elizabeth Baigent. *The Cadastral Map in the Service of the State: A History of Property Mapping.* Chicago, 1992.

Karrow, Robert W. *Mapmakers of the Sixteenth Century and Their Maps.* Chicago, 1993.

Keates, J. S. *Understanding Maps.* London, 1982.

Kendall, D. G. "The Recovery of Structure from Fragmentary Information." *Philosophical Transactions of the Royal Society of London,* ser. A, 279 (1975): 559–75.

Keuning, Johannes. "Sixteenth-Century Cartography in the Netherlands." *Imago Mundi* 9 (1952): 35–63.

Kishimoto, H. *Cartometric Measurements.* Zurich, 1968.

Klingender, Francis D. *Art and the Industrial Revolution.* Frogmore, England, 1972.

Köbel, Jacob. *Geometrei.* Frankfurt, 1536.

Koeman, Cornelis. *Collections of Maps and Atlases in the Netherlands.* Leiden, 1961.

———. *Geschiedenis van de kartografie van Nederland.* Alphen aan den Rijn, 1983.

Konvitz, Joseph. *Cartography in France, 1660–1848.* Chicago, 1987.

Kretschmer, Ingrid, Johannes Dorflinger, and Franz Wawrik. *Lexikon zur Geschichte der Kartographie von den Anfängen bis zum Ersten Weltkrieg.* Vienna, 1986.

Kubelik, Martin. *Die Villa im Veneto.* 2 vols. Munich, 1977.

Laborie, P. J. *The Coffee Planter of Santo Domingo.* London, 1798.

Lago, Luciano. *Theatrum Adriae: Dalle Alpi all'Adriatico nella cartografia del passato.* Trieste, 1989.

Landschaftsverband Rheinland. *Landkarten als Geschichtsquellen.* Cologne, 1985.

Lang, Arend. *Kleine Kartengeschichte Frieslands zwischen Ems und Jade.* Norden, Germany, 1962.

Leerhof, Heiko. *Neidersächsen in alten Karten.* Neumünster, 1985.

———, ed. *Niedersächsen in alten Karten: Eine Ausstellung der Niedersächsischen Archivverwaltung.* Exhibit catalog. Göttingen, 1976.

Lees, William. "The Historical Development of Limerick Plantation." *South Carolina Historical Magazine* 82 (1981): 44–62.

Leigh, Valentine. *The Moste Profitable and Commendable Science, of Surveying of Landes, Tenementes, and Heridamentes.* London, 1577.

Leybourn, William. *The Compleat Surveyor.* London, 1653.

Lingenberg, Heinz. "Zur Geschichte der Kartographie Preussens." *Nord-Ost Archiv* 26–27 (1973): 3–13.

Long, Edward. *The History of Jamaica.* London, 1774.

Lynam, Edward. *The Mapmaker's Art: Essays on the History of Maps.* London, 1953.

Maguire, W. A. *The Downshire Estate in Ireland, 1801–1845: The Management of Irish Landed Estates in the Early Nineteenth Century.* Oxford, 1972.

Maitland, F. W. *Township and Borough.* Cambridge, 1898.

Maling, D. H. "How Long Is a Piece of String?" *Cartographic Journal* 5 (1968): 147–56.

Manley, G. "The Earliest Extant Map of the County of Durham." *Transactions of the Architectural and Archaeological Society of Durham and Northumberland* 7 (1936): 278–87.

Marsala, Vincent John. *Sir John Peter Grant: Governor of Jamaica, 1866–1874.* Kingston, 1972.

Martin, Lawrence, ed. *The George Washington Atlas.* Washington, D.C., 1932.

Martullo Arpago, M. A., LiCastaldo Manfredonia, I. Principe, and V. Valerio, eds. *Fonti Cartografiche nell'Archivo di Stato di Napoli.* Naples, 1987.

Mason, A. Stuart. *Essex on the Map: The Eighteenth-Century Land Surveyors of Essex.* Chelmsford, 1990.

Meinig, Donald. *The Shaping of America. Vol. 1. Atlantic America, 1492–1800.* New Haven, 1986.

Merrens, H. Roy. *Colonial North Carolina in the Eighteenth Century.* Chapel Hill, 1964.

Mertens, Jurgen. *Die neuere Geschichte der Stadt Braunschweig in Karten, Plänen, und Ansichten.* Exhibit catalog. Braunschweig, 1981.

Meurer, Peter. *Fontes cartographici orteliani.* Weinheim, 1991.

Mintz, Sidney W., and Richard Price. *An Anthropological Approach to the Afro-American Past: A Caribbean Perspective.* Philadelphia, 1976.

Moreno Fraginals, Manuel. *El ingenio: Complejo economico social cubano del azúcar.* 2 vols. Havana, 1978.

Morgan, Philip. "The Development of Slave Culture in Eighteenth-Century Plantation America." Ph.D. thesis, University College, London, 1977.

Mosselmans, Jean. *Les Géomètres-Arpenteurs du XVIe au XVIIIe siècle dans nos provinces.* Brussels, 1976.

Mowat, J. L. G., ed. *Sixteen Old Maps of Properties in Oxfordshire in the Possession of the Colleges of Oxford, Illustrating the Open Field System.* Oxford, 1888.

Mundy, Barbara. "The Maps of the *Relaciones Geográficas*

of New Spain, 1579–c. 1584." Ph.D. thesis, Yale University, 1993.

Murphy, Joan. "Measures of Map Accuracy Measurement and Some Early Ulster Maps." *Irish Geography* 11 (1978): 88–101.

Musculus, Johann Conrad. *Der Deichatlas des Johann Conrad Musculus von 1625/26.* Ed. Albrecht Eckhardt. Oldenburg, 1985.

Norden, John. *The Surveyors Dialogue.* London, 1618.

Oehme, Ruthardt. *Die Geschichte der Kartographie des deutschen Südwestens.* Konstanz and Stuttgart, 1961.

———. "Johann Andreas Rauch and His Plan of Rickenbach." *Imago Mundi* 9 (1967): 104–7.

———. *Johannes Oettinger, 1577–1633: Geograph, Kartograph, und Geodät.* Stuttgart, 1982.

Pannett, D. "The Manuscript Maps of Warwickshire, 1597–1880." *Warwickshire History* 6, no. 3 (1985): 69–85.

Panofsky, Erwin. *Albrecht Dürer.* 2 vols. London, 1958.

Parry, M. L., and T. R. Slater, eds. *The Making of the Scottish Countryside.* London, 1980.

Patterson, Orlando. *The Sociology of Slavery.* London, 1967.

Pelletier, Monique. "De nouveaux plans de forêts à la Bibliothèque Nationale." *Revue de la Bibliothèque Nationale* 29 (1988): 56–62.

Pett-Conklin, Linda-Marie. "Cadastral Surveying in Colonial South Carolina: A Historical Geography." Ph.D. thesis, Louisiana State University, 1986.

Pfinzing, Paul. *Methodus geometrica.* Nuremberg, 1598.

Phillips, Ulrich Bonnell. *Life and Labor in the Old South.* Boston, 1929.

Pitz, Ernst. *Landeskulturtechnik, Markscheide, und Vermessungswesen im Herzogtum Braunschweig bis zum Ende des 18. Jahrhunderts.* Göttingen, 1967.

Postgate, M. R. "Field Systems of Cambridgeshire." Ph.D. thesis, University of Cambridge, 1964.

Price, D. J. "Medieval Land Surveying and Topographical Maps." *Geographical Journal* 12 (1955): 1–10.

Prieur, Jutta, and Gerhard Aymans. *Handgezeichnete Karten im Stadtarchiv Wesel.* Wesel, 1987.

Pühler, Christoff. *Geometria oder Feldmessung.* N.p., 1563.

Puppi, Lionello. *Andrea Palladio.* [Venice?], n.d.

Quaini, Massimo, ed. *Carte e cartografi in Liguria.* Genova, 1986.

Ravenhill, W. "The Plottes of Morden Mylles, Cuttell (Cotehele)." *Devon and Cornwall Notes and Queries* 35, pt. 5 (spring 1984): 165–74, 182–83.

Reinartz, Manfred. *Villingen-Schwenningen und Umgebung in alten Karten und Plänen.* Exhibit catalog. Villingen-Schwenningen, 1987.

Reinhold, Erasmus. *Bericht vom Feldmessen.* Erfurt, 1574.

Richeson, A. W. *English Land Measuring to 1800: Instruments and Practices.* Cambridge, Mass., 1966.

Ristow, Walter W., ed. *A la Carte: Selected Papers on Maps and Atlases.* Washington, D.C., 1972.

Robinson, Arthur H., and Barbara Bartz Petchenik. *The Nature of Maps.* Chicago, 1976.

Römer, Gerhard, ed. *Der Neckar in alten Landkarten.* Karlsruhe, 1988.

Rosenau, Helen. *Social Purpose in Architecture: Paris and London Compared, 1760–1800.* London, 1970.

Royal Commission on Historical Monuments, England. *An Inventory of Historical Monuments in the County of Cambridge,* vol. 1, *West Cambridgeshire.* London, 1968.

Schäfer, Karl. "Leben und Werk des Korbacher Kartographen Joist Moers." *Geschichtsblätte für Waldeck* 67 (1979): 123–77.

Schickhardt, Wilhelm. *Kurze Anweisung wie künstliche Landtafeln aus rechtem Grund zu machen.* Tübingen, 1669.

Schnelbögl, Fritz. "The Cartography of Erhard Etzlaub." *Imago Mundi* 20 (1966): 11–26.

———. *Dokumente zur Nürnberger Kartographie.* Nuremberg, 1966.

Schweikher, Heinrich. *Der Atlas des Herzogtums Württemberg vom Jahre 1575.* Ed. Wolfgang Irtenkauf. Stuttgart, 1979.

Schwenter, Daniel. *Geometriae practicae novae.* Nuremberg, 1617.

Seck, Friedrich, ed. *Wilhelm Schickard, 1592–1635: Astronom, Geograph, Orientalist, Erfinder der Rechenmaschine,* Tübingen, 1978.

Seymour, W. A., ed. *A History of the Ordnance Survey.* Folkestone, 1980.

Shadwell, C. L. *The Universities and College Estates Acts, 1858–1880: Their History and Results.* Oxford, 1898.

Shelby, L. R. *John Rogers: Tudor Military Engineer.* Oxford, 1967.

Sheridan, Richard B. *Sugar and Slavery: An Economic History of the British West Indies, 1623–1775.* Barbados, 1974.

Skelton, R. A. *County Atlases of the British Isles, 1579–1703.* London, 1970.

———. *Maps: A Historical Survey of Their Study and Collecting.* Chicago, 1975.

———. "The Military Survey of Scotland, 1747–1755." *Scottish Geographical Magazine* 83 (1967):5–16.

Skelton, R. A., and P. D. A. Harvey, eds. *Local Maps and Plans from Medieval England.* Oxford, 1986.

Skelton, R. A., and J. Summerson. *A Description of Maps and Architectural Drawings in the Collection Made by William Cecil, First Baron Burghley, Now at Hatfield House.* London, 1971.

Société Royale des Bibliophiles et Iconophiles de Belgique. *Les Albums de Croy.* Exhibit catalog. Brussels, 1979.

Souden, David. *Wimpole Hall, Cambridgeshire.* London, 1991.

Steer, F. W. "A Dictionary of Land Surveyors in Britain." *Cartographic Journal* 4 (1967): 124–26.

Stilgoe, John R. *Common Landscape of America, 1580–1845.* New Haven, 1982.

Stöffler, Johannes. *Von künstlicher Abmessung.* Frankfurt, 1536.

Stoney, Samuel Gaillard. *Plantations of the South Carolina Low Country.* New York, 1989.

Strauss, Gerald. *Sixteenth-Century Germany: Its Topography and Topographers.* Madison, 1959.

Strauss, Walter. *The Complete Drawings of Albrecht Dürer.* New York, 1977.

Tait, A. A. *The Landscape Garden in Scotland, 1735–1835.* Edinburgh, 1980.

Taylor, E. G. R. *The Mathematical Practitioners of Hanoverian England.* Cambridge, 1966.

———. *The Mathematical Practitioners of Tudor and Stuart England.* Cambridge, 1954.

Third, Betty M. W. "The Signficance of Scottish Estate Plans and Associated Documents." *Scottish Studies* 1 (1957):39–64.

Thompson, F. M. L. *Chartered Surveyors: The Growth of a Profession.* London, 1968.

Thrower, Norman, ed. *The Compleat Plattmaker.* Berkeley, 1978.

Trewartha, Glenn T. "Types of Rural Settlement in Colonial America." In *Readings in Cultural Geography,* ed. Philip L. Wagner and Marvin W. Mikesell. Chicago, 1962.

Tyacke, Sarah, ed. *English Map-Making, 1500–1650.* London, 1983.

Uzes, François D. *Chaining the Land: A History of Surveying in California.* Fort Sutter Station, Calif., 1977.

Vignoles, K. H. "Charles Blacker Vignoles in South Carolina and Florida, 1817–1823." *South Carolina Historical Magazine* 85 (1984): 83–107.

Vollet, Hans. *Abriss der Kartographie des Fürstentums Kulmbach-Bayreuth.* Exhibit catalog. Kulmbach, 1977.

———. *Weltbild und Kartographie im Hochstift Bamberg.* Kulmbach, 1988.

———, ed. *Oberfranken im Bild alter Karten: Ausstellung des Staatsarchiv Bamberg.* Bamberg, 1983.

Vries, Dirk de. *Kaarten met Geschiedenis, 1550–1800.* Utrecht, 1989.

Waddell, Hope Masterton. *Twenty-nine Years in the West Indies and Central Africa.* London, 1863.

Wagner, Philip, and Marvin Mikesell, eds. *Readings in Cultural Geography.* Chicago, 1962.

Walford, Edward. *The County Families of the United Kingdom.* London, 1860.

Wallis, Helen, and Arthur H. Robinson, eds. *Cartographical Innovations: An Internatinal Handbook of Mapping Terms to 1900.* London, 1987.

Webb, Stephen Saunders. *The Governors-General: The English Army and the Definition of Empire, 1569–1681.* Chapel Hill, N.C., 1979.

Westra, F. "De landmeter-ingenieur Johan Sems en de kaarten van Leewarden (1600–1603) en Franeker." *Caert-Thresoor* 2 (1983): 2–5.

Widekind, Herman. *Bewerte Feldmessung und Theilung.* Heidelberg, 1578.

Wilms, Douglas. "The Development of Rice Culture in Eighteenth-Century Georgia." *Southeastern Geographer* 12 (1972): 45–57.

Winschiers, Kurt. *500 Jahre Vermessung und Karte in Bayern.* Munich, 1982.

Wise, Donald. "The Young Washington as a Surveyor." In *Northern Virginia Heritage.* N.p., 1979.

Wobeser, Gisela von. *La formación de la hacienda en la epoca colonial.* Mexico City, 1983.

Wolff, Hans, ed. *Cartographia Bavariae: Bayern im Bild der Karte.* Catalog. Weissenhorn, 1988.

———. *Philipp Apian und die Kartographie der Renaissance.* Exhibit catalog. Weissenhorn, 1989.

Woolgar, C. M. "Some Draft Estate Maps of the Early Seventeenth Century." *Cartographic Journal* 22 (1985):136–43.

Wright, Philip, ed. *Lady Nugent's Journal.* Kingston, 1966.

Wunderlich, Herbert. *Kursächsische Feldmesskunst, artilleristische Richtwerfahren, und Ballistik im 16. und 17. Jahrhundert.* Berlin, 1977.

Zandvliet, Kees. *The Beesbos near Geertruidenburg by Peter and Jacob Sluyter.* The Hague, 1979.

———. *De Groote Waereld in't Kleen Geschildert.* Alphen aan den Rijn, 1985.

Zögner, Lothar, ed. *Bibliographie zur Geschichte der deutschen Kartographie.* Exhibit catalog. Munich, 1984.

———. *Von Ptolemaus bis Humboldt: Kartenschätze des Staatsbibliothek Preussischer Kulturbesitz.* Exhibit catalog. Berlin, 1985.

Zögner, Lothar, and Gudrun Zögner. *Preussens ämtliche Kartenwerke im 18. und 19. Jahrhundert.* Exhibit catalog. Berlin, 1981.

INDEX